Active Metamaterials

Saroj Rout • Sameer Sonkusale

Active Metamaterials

Terahertz Modulators and Detectors

Springer

Saroj Rout
Mixignal Innovations
Nashua, NH, USA

Sameer Sonkusale
Electrical and Computer Engineering
Tufts University
Medford, MA, USA

ISBN 978-3-319-52218-0 ISBN 978-3-319-52219-7 (eBook)
DOI 10.1007/978-3-319-52219-7

Library of Congress Control Number: 2017930655

Printed on acid-free paper

This Springer imprint is published by Springer Nature
The registered company is Springer International Publishing AG
The registered company address is: Gewerbestrasse 11, 6330 Cham, Switzerland

To our parents

Preface

Metamaterials, first known as left-handed materials (LHMs) or negative-index materials (NIMs), are artificially constructed *effective materials* using periodic structures that are a fraction of the wavelength of the incident electromagnetic wave, resulting in effective electric and magnetic properties (permittivity and permeability) that are unavailable in natural materials. The ability to design effective materials with deterministic electromagnetic properties makes them very attractive for terahertz applications since natural materials do not respond well to this frequency regime.

Since the discovery and experimental demonstration of artificial material with negative permittivity and negative permeability by Sir Pendry et al. in 1996 and 1999, the research interest in metamaterials has significantly increased. In spite of the intense research activities in the last two decades, application of metamaterials to terahertz frequencies is a recent phenomenon. And, terahertz metamaterials embedded with active devices are even a smaller portion of this research landscape. Moreover, the transition from research to real-world application of terahertz metamaterials is still years behind because of numerous implementation challenges.

This book intends to close that gap by providing theoretical background and experimental and fabrication methods in one comprehensive text. This is well suited for engineers and physicists to be able to design, fabricate, and characterize terahertz metamaterial devices in commercial planar semiconductor processes.

Three case studies are covered in detail involving terahertz modulator and detector implemented in commercial gallium arsenide (GaAs) and complementary metal-oxide semiconductor (CMOS) process for imaging and communication applications.

The first three chapters provide the introduction, background theory, and experimental methods which give the reader the motivation and basic background to understand terahertz metamaterials. The last three chapters provide the three case studies of active metamaterials fabricated in planar semiconductor process for terahertz imaging and communication applications.

Chapter 1 begins by providing the motivation for working in the terahertz frequency regime for its numerous important applications and showing, with quantitative reasoning, why metamaterials are a suitable technology for that regime. An overview of technologies for terahertz wave modulators is presented, which is the primary structure underlying all of the designs in this text.

Chapter 2 reviews some of the basic electromagnetic principles for a basic understanding of metamaterials. One of the key contributions of this text is the analysis of terahertz wave modulators using the Drude–Lorentz model.

Chapter 3 covers the experimental methods for modeling, simulating, and characterizing terahertz metamaterials. A section on CMOS fabrication is introduced with few metamaterial case studies for readers to get familiar with a very accessible process that can be used for limited terahertz metamaterial applications. For metamaterial characterization, terahertz time-domain spectroscopy (THz-TDS) and continuous-wave terahertz spectroscopy (cw-THz) are covered in detail. A section is also dedicated to the alignment of off-axis parabolic mirrors, which is found in most terahertz test setups and should be a valuable resource for anyone doing experiments in this field.

Chapter 4 covers a case study of metamaterial-based terahertz modulator using embedded HEMT devices which is one of the main contributions to the scientific literature by the authors. Design principle of HEMT-controlled metamaterial is covered in detail based on the modulator principle introduced in Chap. 2. Design, fabrication, experimental setup, and test results are also covered in this chapter.

Chapter 5 covers another case study of an all solid-state metamaterial-based terahertz spatial light modulator (SLM) using the HEMT-based modulator described in Chap. 4. The principle behind single-pixel imaging is presented in this chapter followed by design, fabrication, and test of the SLM.

Chapter 6 presents the last case study of a terahertz focal plane array (FPA) using metamaterials in a 0.18 μm CMOS process. The principle of resistive self-fixing detection is covered followed by the design and simulation of the FPA.

Nashua, NH, USA Saroj Rout
Medford, MA, USA Sameer Sonkusale

Acknowledgments

There have been quite a few who have inspired and motivated this work and helped us design, simulate, and characterize. We want to begin by thanking Dr. Willie Padilla whose department colloquiums at Tufts University were to sow the seeds for the contributions we made in this text. We want to thank him for introducing us to this exciting field of metamaterials. We also want to thank his doctoral students Dr. David Shrekenhamer and Chris Bingham for help with the design and simulation of the modulators. David in particular spent countless hours helping us simulate and test the HEMT-based metamaterial modulator we developed. We also want to thank Dr. Richard Averitt and Dr. Andrew Strikwerda for helping us with their terahertz time-domain spectroscopy (THz-TDS) setup for characterizing the modulator.

We thank Dr. Wangren Xu for contributing Sect. 3.2 on design for fabrication in foundry process. Our thanks to Jessie Tovera for assembly of the spatial light modulator (SLM).

Our special thanks to Mr. Charles B. Glaser, editorial director at Springer USA, for encouraging us with writing the proposal of this book and providing us feedback and clarification throughout the writing process.

We acknowledge the support of the Office of Naval Research under US Navy Contracts N00014-07-1-0819 and N00014-09-1-1075, and the National Science Foundation Awards No. ECCS-1002340 and ECCS-1002152.

We also acknowledge the support of DoD Defense University Research Infrastructure Program (DURIP) under grant N00014-12-1-0888 equipment purchase used for test and characterization.

Finally, we want to thank our families for their unconditional love and support without which nothing would ever be possible.

Contents

Chapter 1
Introduction

The primary purpose of this text is to provide analytic and experimental methods to design planar metamaterial based wave modulators for terahertz frequencies, especially for communication and imaging applications. This chapter provides the motivation for the reader by answering two basic questions: why terahertz frequencies and why design metamaterials for that frequency regime.

Section 1.1 introduces the fact behind the so-called terahertz gap and illustrates numerous important examples to show why this frequency regime is interesting. A brief history of various terahertz technologies is provided for the reader to get a historical perspective as well as the state of the art. Section 1.2 introduces the basic concept of metamaterials with a brief history and overview of the topic. A quantitative reason is provided in Sect. 1.2.3 as to why metamaterials are suitable for terahertz frequencies. Section 1.3 provides an overview of technologies for terahertz wave modulators, which is the primary focus of this text.

1.1 Towards Closing the "Terahertz Gap"

The terahertz (THz) regime of the electromagnetic (EM) spectrum is broadly recognized by the frequency range of 100 GHz to 10 THz[1] (where 1 THz corresponds to a frequency of 10^{12} Hz, a wavelength of 300 μm, and photon energy of 4.1 meV) [2, 3]. This region, alternatively called the far-IR, lies between infrared light and the microwave frequencies as shown in Fig. 1.1a.

Most modern devices are based on the response of the electrons in materials to applied EM field and the nature of the response is dependent on the frequency of the EM field. At frequencies few hundred gigahertz and lower, the motion of free electrons forms the basis of most EM devices characterized broadly as *electronics*.

[1]Some define THz to be the submillimeter-wave energy that fills the wavelength range between 1 and 0.1 mm (300 GHz–3 THz) [1].

S. Rout, S. Sonkusale, *Active Metamaterials*, DOI 10.1007/978-3-319-52219-7_1

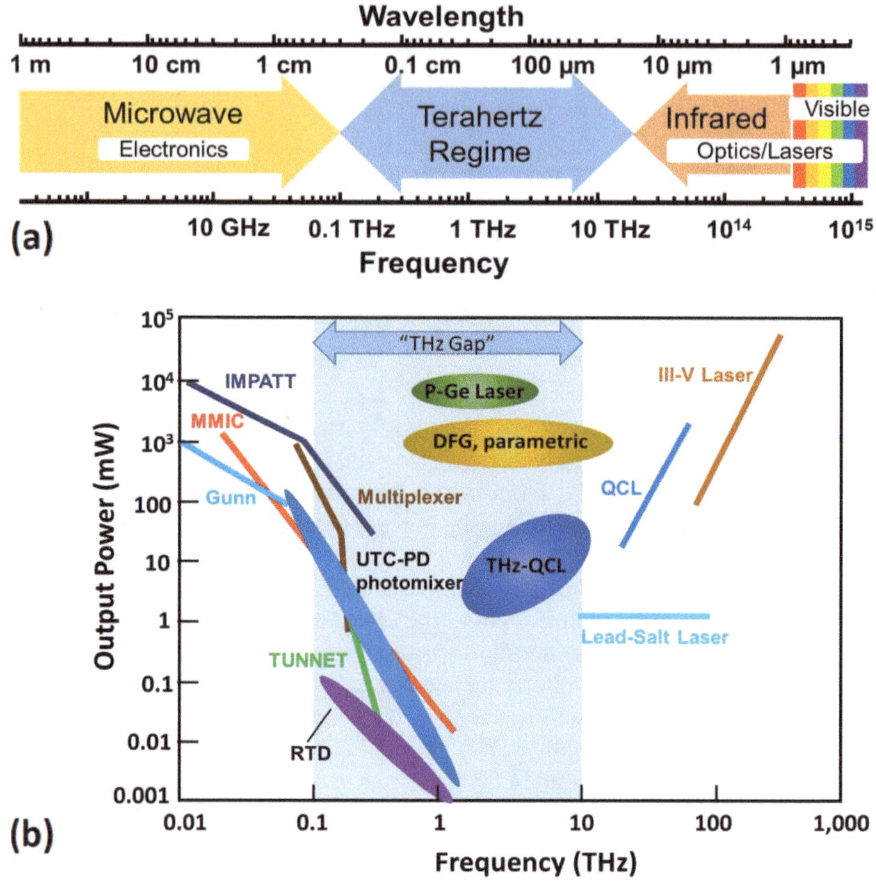

Fig. 1.1 The THz gap—(**a**) THz regime of the EM spectrum extends from 100 GHz to 10 THz, which lies below visible and infrared (IR) wavelengths and above microwave wavelengths. (**b**) Illustration of the THz gap—THz emission power as a function of frequency [4]. *Solid lines* are conventional sources and *oval shapes* denote recent THz sources. IMPATT stands for impact ionization avalanche transit-time diode, MMIC stands for microwave monolithic integrated circuit, TUNNET stands for tunnel injection transit time, multiplexer is a Schottky barrier diode (SBD) frequency multiplier, UTC-PD stands for uni-traveling-carrier photodiode, DFG stands for difference frequency generator and QCL stands for quantum cascade laser

It is well known that fundamental sources of power in electronics exhibit decrease in power with increasing frequency. This is a consequence of few main factors such as characteristic transit distance, for example related to length of a channel in a FET; second is the saturation velocity of carriers in the device, typically 10^5 m/s; and the third being the maximum electric field that is sustainable in a device before breakdown. The accumulation of these limits has led to practical electronic RF sources around 300 GHz with sub-milliwatt power output [5] and much lower power for frequencies near THz, for example, 15 μW at 1.8 THz for a Schottky multiplier chain [6].

On the other hand, at infrared through optical and ultra-violet wavelengths the devices are characterized as *photonics*. It is also well known that photonics sources exhibit decrease in output power with decreasing frequency and the fundamental reason for this is lack of photon energy to meet the band-gap requirements for most photonics sources: for example, the longest wavelength lead salt laser diodes do not extend below 15 THz. Thus, in between these two regions, there exists the so-called THz gap, where the efficiency of electronic and photonics devices tends to taper off as illustrated in Fig. 1.1b. Because of the lack of natural materials that respond to the THz region of EM spectrum, this regime is arguably the least developed and least understood of the EM spectrum [2, 4, 5].

1.1.1 Why Is the "Terahertz Gap" Interesting

The past three decades have seen a revolution in THz systems motivated in part by vast range of unique applications ranging from imaging, sensing to spectroscopy [7], as shown in Fig. 1.2. Astronomy and space research has been one of the early drivers for THz research because of the vast amount of spectral information available concerning the presence of abundant molecules such as oxygen, water, and carbon monoxide in stellar dusts, comets, and planets that have unique THz spectral signatures [8]. The application of THz sources in astronomy is as local oscillator sources for submillimeter-wave heterodyne receivers used to perform high-resolution spectroscopy [1]. Back on Earth, the two most pervasive applications for

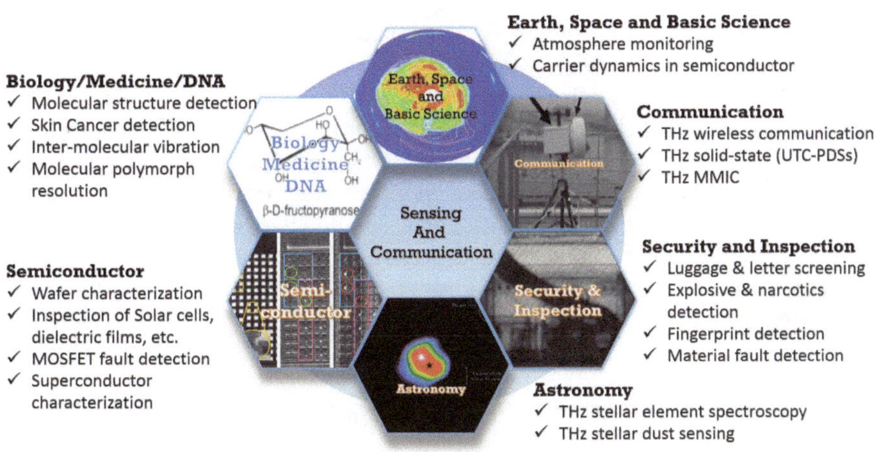

Fig. 1.2 A snapshot of some of the unique applications in the terahertz frequency regime

terahertz technology have been in the areas of plasma fusion diagnostics and gas spectroscopy [9]. THz spectroscopy development has been an active research topic due to its potential in extracting material characteristics that are unavailable when using other frequency bands. Terahertz spectroscopy has been used to determine the carrier concentration and mobility of doped semiconductors such as GaAs and silicon wafers [10, 11] and high-temperature superconductor characterization [12]. THz systems have been developed for biomedical applications that are facilitated by the fact that the collective vibrational modes of many proteins and DNA molecules are predicted to occur in the THz range [13]. A further biomedical application of THz systems is the THz biosensor. A simple biosensor has been demonstrated for detecting the glycoprotein avidin after binding with vitamin H (biotin) using the technique of differential THz-TDS [14].

In recent years, terahertz (THz) imaging has captured significant interest due to its ability to penetrate most dielectric materials and non-polar liquids [15, 16], allowing stand-off spectroscopy for security screening [17] and illicit drug detection [18]. Its harmless interaction with human tissue has resulted in wide range of applications such as bio-detection [19] and skin cancer detection [20], to name a few. Another attraction of THz imaging is largely due to the availability of phase-sensitive spectroscopic images, which holds the potential for material identification or "functional imaging." THz systems are ideal for imaging dry dielectric substances including paper, plastics and ceramics. These materials are relatively non-absorbing in this frequency range, yet different materials may be easily discriminated on the basis of their refractive index, which is extracted from the THz phase information. Many such materials are opaque at optical frequencies, and provide very low contrast for X-rays. THz imaging systems may therefore find important niche applications in security screening and manufacturing quality control [21]. Interest in using THz imaging to study cellular structure is also increasing [22].

Another emerging area for the terahertz regime is high-speed wireless communication. The demand for high-speed wireless access is increasing due to large amounts of data needed for new emerging applications such as the consumer market that is already demanding 20, 40 and 100 Gbit/s wireless technologies for Super Hi-Vision (SHV) and Ultra High-Def (UHD) TV data [23]. Terahertz (THz) carrier frequencies will offer the advantage of higher data speed, sub-millimeter antenna size and short range security especially suitable for portable devices. Although they are susceptible to atmospheric loss, THz digital communication systems have been demonstrated near certain windows, especially around the 300–400 GHz range that has shown promise for high bit-rate data transmission [23–25].

Next we show two examples of terahertz systems to highlight the significance of this frequency regime.

1.1.1.1 Continuous-Wave Terahertz System for Inspection Applications

The unique property of terahertz waves to penetrate most dielectric materials allows for stand-off spectroscopy of dangerous substances making them an ideal candidate for security screening applications. One such system, a compact continuous-wave

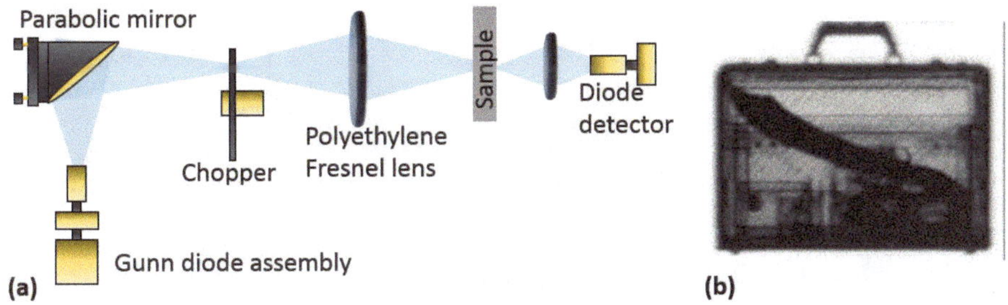

Fig. 1.3 Continuous-wave (cw) terahertz system for inspection applications. (**a**) Schematic diagram of the imaging system in transmission geometry. (**b**) Image of a briefcase holding a large knife and various harmless contents such as a compact disc, a video cassette, and audio cassette and pens [17]

(cw) 0.2 THz imaging system using entirely electronic generation and detection with a minimal number of components is shown in Fig. 1.3a [17]. The system consists of a frequency-doubled Gunn diode oscillator with an output power of 12 mW as a source, and a Schottky diode as the detector. After being emitted by the Gunn diode, the beam is focused by an parabolic mirror to a 4 mm spot where it is modulated by an optical chopper at 1.2 kHz. The chopped beam is focused twice, one on the sample and again on the diode detector by a pair THz optics. The sample is mounted on a translation stage to raster scan the object.

Figure 1.3b shows a 0.2 THz cw scan of a standard size leather briefcase, containing benign and suspicious items. The strengths of the system can be immediately seen: objects can be recognized fairly easily, and since the radiation involved is non-ionizing, it poses little threat to human beings.

1.1.1.2 Giga-Bit Wireless Link Using 300–400 GHz Bands

Because of demand for high-speed wireless access, there is an urgency for designing wireless systems with data speeds of more than 1 Gbit/s. Figure 1.4 shows a block diagram of a 300–400 GHz band system using a photonics-based transmitter and receiver [26]. This system is intended for use in short-distance ($\sim$0.5 m) applications. An optical RF signal is generated by heterodyning the two wavelengths of light from the wavelength-tunable light sources. The optical signal is digitally modulated by a optical intensity modulator driven by a pulse pattern generator (PPG). Finally, the optical signal is converted to an electrical signal by the modified Uni-Traveling-Carrier-Photodiode (UTC-PD). The THz wave is emitted to free space via a horn antenna with a gain of 25 dBi, and it is collimated by a 2-inch-diameter Teflon lens. The receiver consists of a Schottky barrier diode and an IF filter followed by a low-noise pre-amplifier and a limiting amplifier. The envelope detection is performed by the Schottky-Barrier-Diode (SBD) for amplitude shift keying (ASK) modulation. The inset in Fig. 1.4 shows the eye diagram at 14 Gbit/s which is evidently error-free from the clear eye opening.

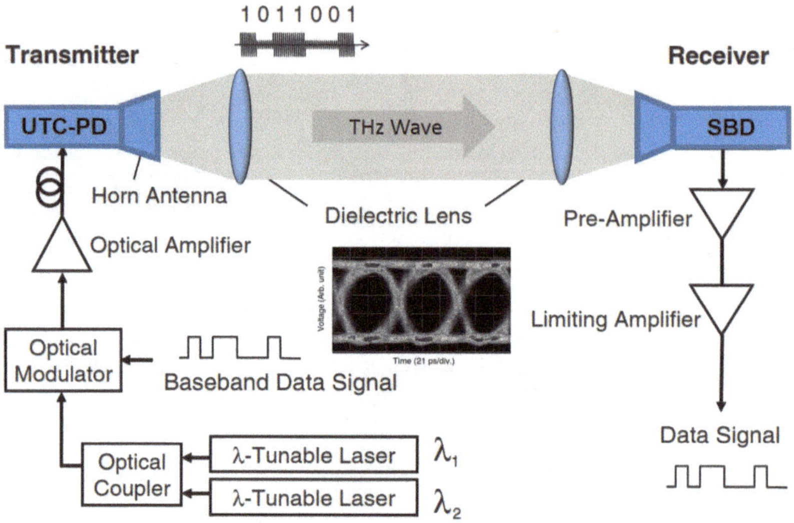

Fig. 1.4 Giga-bit wireless link using 300–400 GHz bands. Block diagram of the wireless link using photonics-based terahertz-wave transmitter. UTC-PD: Uni-Traveling-Carrier-Photodiode. SBD: Schottky-Barrier-Diode. (inset) Eye diagram at 14 Gbit/s [26]

With 14 Gbit/s data rate, it is clear that we need terahertz carrier frequency for the demand we have in hand and these examples are promising for the next generation wireless technology.

1.1.2 A Brief History of Terahertz Technologies

For terahertz imaging, sources and detectors have been the focus of research since the 1960s and the early 1970s. A powerful source developed for terahertz radiation was the HCN laser operating at 1.12 THz [27]. Terahertz imaging got its first boost from the development of far-infrared gas laser and the schottky diode harmonic mixer in the mid-1970s by Hartwick et al. [28]. At roughly the same time, the advent of lasers motivated research in far-infrared generation using the difference-frequency generation in non-linear crystals [29], which served as the seed for the development of terahertz time-domain spectroscopy (THz-TDS) [30–34], which still serves as a system of choice for high dynamic range THz spectroscopy. The THz-TDS system generated a great deal of interest in THz imaging after first images acquired using the THz-TDS was reported in 1995 [35]. Since the early days, the field has seen a rapid expansion in new technologies for terahertz sensing [1] including time-of-flight imaging [36], variety of different tomographic and synthetic aperture imaging [37–39], multi-element detector schemes, both microbolometer arrays [40] and electro-optic sampling with high-performance CCD cameras [41].

The semiconductor-laser model for terahertz generation has always been a topic of research interest, given their tremendous success in visible and near-infrared frequency range. Because of the lack of appropriate interband materials, in the past, artificially engineered materials to obtain the desired transitions has been natural route of pursuit for many years by researchers to create terahertz devices. This pursuit led to the demonstration of the first terahertz quantum-cascade (QC) laser in 1988 [42] and shorter wavelength of 4 μm (75 THz) in 1994 [43]. It was only in October of 2001 that the first QC laser with a photon energy less than the semiconductor optical phonon energy was demonstrated at 4.4 THz (67 μm) [44]. Since then, significant research has led to QC lasers demonstrating spectral coverage from 0.84 to 5.0 THz [45]. One of the major limitations that has prevented QC lasers from being a mainstream technology is the inability to deliver THz power at room temperature. But recent work in 2012 has demonstrated terahertz QC laser operation up to 200 K [46] and a recent work by Li et al. in 2014 demonstrating powers >1 W [47], showing a bright future for QC lasers and maybe the technology to fill the "THz gap." The success of QC lasers is already evident from their wide range applications: such as local oscillator sources for submillimeter-wave heterodyne receivers used to perform high-resolution spectroscopy in astronomy [1, 48] and in many high-performance imaging applications [49–52], including a real-time, video-rate terahertz imaging application [45].

Although short range communication using THz waves was first speculated by Gebbie in 1970 [53], we are still far from realizing systems for our consumer wireless needs due to enormous challenges associated with the technology and the medium as well. Due to the increasing demand for high-speed wireless communication in the last decade, we are starting to see communication systems developed for the THz regime. A photonics-based 120 GHz system has been successfully demonstrated multi-channel transmission of uncompressed high-definition TV signal over a distance of 100 m [54, 55]. With progress in the semiconductor IC technologies, all solid-state THz communication systems are being developed which contributes to reduction in size, weight, and cost together with ease of operation [56]. A trial use of this system was conducted within the live broadcast of the 2008 Olympic Games in Beijing by Nippon Telegraph and Telephone Corporation (NTT) and Fuji Television Network Inc. Towards 20+ Gbit/s wireless systems using the 300–400 GHz band has been experimentally demonstrated using a photonics-based THz system over a short range, suggesting possible utilization of the band for multi-channel gigabit links [26]. Another unique advantage with THz wireless communication is its secure at a physical level for several reasons: highly directional beams compared to microwave communications, less scattering of radiation compared to IR wireless, limited propagation distance due to atmospheric attenuation, encryption of the beam, large channel bandwidth for spread spectrum techniques which enable anti-jamming and low probability of detection systems, and hidden THz signals in the background noise.

1.2 Introduction to Metamaterials

1.2.1 A Brief History

The first published work on "artificial materials" has been traced back to as early as 1898 [57], when Jagadis Chunder Bose demonstrated polarization of microwave electric field using twisted jute [58]. In 1914, Lindman demonstrated "artificial" chiral materials for microwave frequencies [59]. Bose and Lindman were way ahead of their times and can be safely considered pioneers in the field of artificial materials. The earliest demonstration of negative refractive index has been traced to Prof. Mandlshtam's lecture notes at Moscow University in 1945 [60]. In 1948, Kock demonstrated tailoring effective refractive index of media by subwavelenth metallic inclusions [61]. Since then, negative refraction and backward-wave media has been discovered theoretically and experimentally by several researchers [62–64].

It wasn't till 1967, Veselago's seminal work on simultaneous negative values of permittivity(ϵ) and permeability(μ) [65],[2] a systematic analysis of electromagnetic behaviors in left-handed materials (LHM) was proposed, predicting some exotic EM behaviors from them. Although some of these behaviors like negative refraction and backward-wave media had been discovered theoretically in the earlier works, Veselago also showed some new features like reversed Doppler shift and backward Cerenkov radiation in LHM.

Like his predecessors, Veselago's work was ahead of time and went literally unnoticed for three decades due to lack of natural LHM and lack of experimental data as well. In 1996, Pendry et al. demonstrated negative permittivity (ϵ) by realizing artificial plasma using thin wire medium [66], and then in 1999 discovered negative permeability (μ) using the well-known split-ring resonators (SRR) as magnetic elements [67]. By combining both negative ϵ and μ, Smith et al. demonstrated the first negative refractive index artificial material in 2000 [68].

Since then, metamaterial research has exponentially grown and expanded into fields previously un-imagined by the pioneers themselves. This includes negative refractive index [69–73], super-lensing [74–77], perfect absorbers [78, 79], cloaking [80–83], and more generally, coordinating transformation materials [84–87].

During the same time as the research in artificial effective medium for LHM was progressing, an alternative representation of LHM was being researched by three groups (Eleftheriades, Oliner, and Caloz-Itoh), using the transmission-line (TL) approach [88–91]. A conventional transmission line, a series inductance (L) and shunt capacitance (C), can be considered as a one-dimensional (1D) RHM and dual to that is a series C and shunt L that supports backward wave and hence can represent LHM. This new way of looking at traditional transmission line theory has led to a lot of exotic microwave components and antennas [91].

[2]Original Russian version was published in 1967.

Bulk of the research has been focused on passive metamaterials, sub-wavelength metallic inclusions in dielectric material, allowing for a pre-determined resonant response of the electrical permittivity or magnetic permeability. Although passive metamaterials have given us the power to engineer material properties of our choice, they cannot be changed dynamically. For many potential applications, it would be desirable to control these properties through means of electrical, optical, mechanical, temperature methods or any combinations of them.

In 2006, for the first time, dynamical control of the electrical response of the split-ring resonators (SRRs), a basic building block of metamaterials, was demonstrated by Padilla et al. [92], through photo-excitation of free carriers in the substrate. Later that year, Chen et al. [151] also demonstrated electrical control of terahertz metamaterials using dynamic substrate bias voltage. Since then, research into active and tunable metamaterials has grown and expanded into fields previously unrealized using passive metamaterials. This includes photo-controlled [93, 94], electrical controlled [95, 96], temperature controlled [97], MEMS controlled [98] metamaterials. Although these dynamically controlled metamaterials have expanded the scope of application, they are still limited by slow response to the control method. It is due to the fact that the entire substrate of the metamaterial is controlled resulting in large device capacitance [95] in case of electrically controlled metamaterials and similar limitations in other control methods. To alleviate some of these limitations and expand into the next frontier of metamaterial research, it is required to embed and interface solid-state devices locally into every metamaterial unit.

In 2011, our collaborative work [99] demonstrated the first embedded high electron mobility transistor (HEMT) devices in metamaterials to create a high-speed THz modulator. Now we are witnessing another metamaterial frontier, where solid-state devices are embedded in metamaterial units, resulting in even more exotic devices previously not realized.

1.2.2 Overview of Metamaterials

Typically, isotropic materials can be characterized by their effective dielectric (permittivity, ϵ) and magnetic (permeability μ) properties. A ubiquitous material in nature is free space or air, with a permittivity of ϵ_0 and permeability of μ_0. The relative permittivity and permeability of a material are defined as $\epsilon_r = \epsilon/\epsilon_0$ and $\mu_r = \mu/\mu_0$, respectively, from which the refractive index of that material can be written as $n = \sqrt{\epsilon_r \mu_r}$. Figure 1.5 illustrates the classification of isotropic materials in the $\epsilon - \mu$ space. In Fig. 1.5, the first quadrant ($\epsilon > 0$ and $\mu > 0$) represents right-handed materials (RHM), which support the right-handed (forward) propagating waves in most dielectric or optical materials. From Maxwell's equation, the electric field **E**, the magnetic field **H**, and the wave vector **k** form a right-handed triplet. The second quadrant ($\epsilon < 0$ and $\mu > 0$) represents the electric plasma where the incident EM wave decays and supports evanescent waves. Many metals in the ultra-violet and visible frequency range fall in this quadrant. The third quadrant

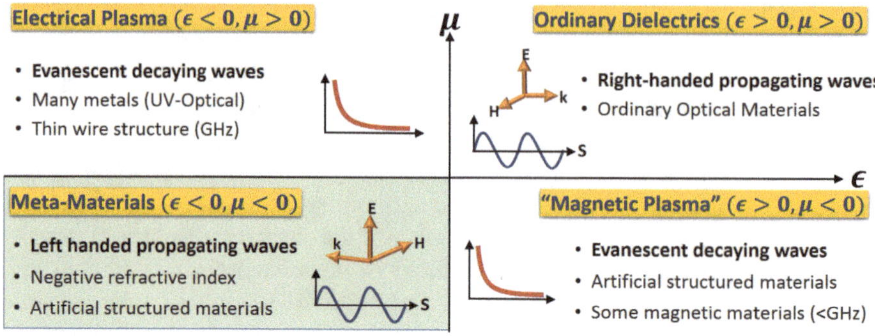

Fig. 1.5 Classification of materials based on dielectric (ϵ) and magnetic (μ) properties. First quadrant represents materials with $\epsilon > 0$ and $\mu > 0$, the second represents $\epsilon < 0$ and $\mu > 0$, the third denotes $\epsilon < 0$ and $\mu < 0$ and the fourth represents $\epsilon > 0$ and $\mu < 0$

($\epsilon < 0$ and $\mu < 0$) represents the left-handed material (LHM) which supports many exotic electromagnetic properties, such as backward propagating waves, as predicted by Veselago in 1968 [65]. There are no known natural materials that exhibit properties of this quadrant. In LHM, the $\mathbf{E} - \mathbf{H} - \mathbf{k}$ triplet is given by the left-hand rule. The fourth quadrant ($\epsilon > 0$ and $\mu > 0$) denotes the *magnetic* plasma, which supports evanescent waves and very few natural ferromagnetic materials at sub-GHz frequencies fall in this category.

Metamaterials are macroscopic composite of periodic or non-periodic sub-wavelength structures, whose EM property is primarily a function of the cellular structure allowing great flexibility in creating new effective materials, unavailable in nature. The sub-wavelength inclusions are like the atomic particle in conventional material and the EM response of the metamaterial is characterized by the effective permittivity(ϵ) and permeability(μ), determined in the homogenization of the periodic inclusions by averaging the local fields [100, 101]. In the early days, the term metamaterial referred to negative-refractive index material (NIM), the third quadrant in Fig. 1.5, phenomena first postulated by Veselago in 1968 [65]. He theoretically predicted materials with simultaneous negative values of the ϵ and μ, hence giving an index of refraction, $n = \sqrt{\epsilon_r \mu_r}$ that is less than zero, a material not naturally found even today. Along with negative-refractive index, Veselago also predicted other exotic properties such as reverse Cerenkov radiation, reverse Doppler shift, and opposite phase and group velocity, among others [65]. In its long history, metamaterials, Left-Handed materials (LHM), NIM, double-negative materials (DNG), and backward-wave materials have been regarded as the same terms. But today, the term metamaterial has a much broader scope than LHM. They constitute any artificial EM materials constructed from sub-wavelength periodic or non-periodic inclusions resulting in a user defined effective ϵ, μ or both. In terms of the classification in Fig. 1.5, today metamaterials denote any artificial effective medium that represents quadrant two, three and four.

Design of negative-index metamaterial (NI MM) constitutes magnetic and electric elements that, respectively, give rise to negative ϵ and negative μ. Various types of sub-wavelength resonators for building MMs have been designed and demonstrated, for example, thin metallic wires [66, 102], Swiss rolls [103, 104], pairs of rods and crosses [105, 106], fishnet structures [107, 108] and split-ring resonators (SRRs) and electric variants of it [68, 79, 98–100, 104, 109–112, 151]. Among these resonator structures, the SRRs and their variants are the canonical sub-wavelength structures used in majority of the metamaterials. In the next two sections, we will discuss design techniques to realize negative ϵ or negative μ metamaterials.

1.2.2.1 Magnetic Split-Ring Resonator (SRR)

The first artificial magnetic material was demonstrated in the microwave frequency range by Pendry et al. in 1999 [104] using a "magnetic atom" very similar to a frequently used split-ring resonators (SRRs) arranged in an array as shown in Fig. 1.6a. The SRR array is built from non-magnetic conducting sheets on an

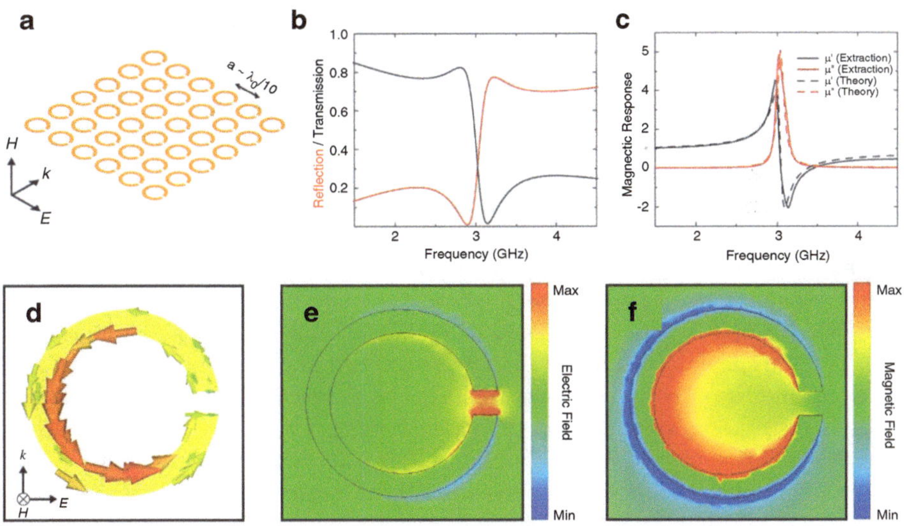

Fig. 1.6 Magnetic split-ring resonators (SRRs). (**a**) Incident electromagnetic wave propagates across the SRR array, where the element spacing a is 1/10 of the free space wavelength λ_0. (**b**) The simulated reflection and transmission for an SRR with $a = 10\,\text{mm}$. (**c**) The extracted magnetic response (*solid curves*) from simulation results shown in (**b**) and the response when fit with the theoretical model listed in Eq. (1.1), where $F = 0.20$, $\omega_0 = 2\pi \times 3.03\,\text{GHz}$, and $\gamma = 0.10\,\text{s}^{-1}$. (**d**)–(**f**) Resonant field distributions in the plane of the metamaterial where (**d**) shows the resonant current density (**e**) the maximum electric field, and (**f**) plots maximum magnetic field [113]

FR4 dielectric whose effective material response is shown in Fig. 1.6b, c that is determined solely by the SRR geometry. The general form of the frequency dependent permeability of the SRR has the generic form [104, 114]

$$\mu_{\text{eff}}(\omega) = 1 + \frac{F\omega^2}{\omega_0^2 - \omega^2 - i\gamma\omega} \tag{1.1}$$

$$= \mu'_{\text{eff}} + i\mu''_{\text{eff}} \tag{1.2}$$

where F is a geometrical factor, ω_0 is the resonance frequency, and γ is the resistive damping factor. The quality factor of the resonator can be expressed as $Q = \omega_0/\gamma$ where γ is the bandwidth of the resonator. The Q is related inversely to the resistive damping factor (γ). If the Q of the resonator is high enough, the *real* part of Eq. (1.1) will result in a negative value around the resonant frequency, yielding an effective negative magnetic response from the SRR array.

In its simplest form, the SRR can be thought of as an inductor-capacitor (LC) resonator, with a resonance frequency of $\omega_0 \approx \sqrt{1/LC}$, where the inductance results from the circular current path in the split ring and the capacitance from the dielectric in the split gap. When a time varying magnetic field polarized perpendicular to the plane of SRR is incident on the material as shown in Fig. 1.6a, it induces a circulating current according to Faraday's law that causes charges of opposite polarity to accumulate at the split gap of the split ring, thereby producing a strong enhancement of electric field as shown in Fig. 1.6e. For frequencies below ω_0, the current in the loop and the incident field are in phase, resulting in a positive response. However, when the frequency of the **B**-field is in the proximity of the resonant frequency, ω_0, the currents in the SRR lag thus resulting in a negative response. Figure 1.6d–f shows the resonant current, electric, and magnetic field distributions, respectively, for the excited SRR.

The LC resonator model is very simple but powerful design tool for achieving an initial resonant response very close to the desired frequency by simply scaling these variables. In practice, other parameters such as frequency dependent dielectrics and scattering effects will affect the resonant frequency. EM simulators are used to iterate and fine-tune the resonant frequency to the desired one. This description is also helpful for designing dynamic metamaterials, wherein majority of research focuses on different mechanisms of controlling either the capacitance associated with the split gap, the length of the resonator or the corresponding resistivity loss.

A potential limitation of SRRs to be used for planar THz devices is that the magnetic field needs to be perpendicular to the SRR plane for full magnetic coupling. However the EM plane waves are usually incident normal to the planar SRR structure with the magnetic field lying in the SRRs plane, which does not excite the magnetic resonance directly. Variants of the SRR which respond to the electric field instead, allowing normal incident of the plane wave, is introduced in the next section.

1.2.2.2 Electrically Coupled LC Resonator (ELC)

Researchers have been constructing artificial EM materials with $\epsilon < 0$ (artificial plasmas) for decades now [115–118]. The technology has been reintroduced in recent years with more physics-oriented understanding [66, 119]. The first artificial electric material was demonstrated in the microwave frequency range by Pendry et al. [66], using the straight wire medium, a principle first demonstrated by Rotman in 1962 [116]. The generic form of the frequency dependent permittivity, in naturally occurring materials, can be written as a Drude–Lorentz relation:

$$\epsilon_{\text{eff}}(\omega) = 1 - \frac{\omega_p^2}{\omega^2 - \omega_0^2 - i\gamma\omega} \tag{1.3}$$

where plasma frequency, ω_p^2, is

$$\omega_p^2 = \frac{4\pi n e^2}{m^*} \tag{1.4}$$

and n is the carrier density, e is the charge of an electron, γ is the damping factor and m^* is the effective mass of carriers. In artificially EM materials, including MMs, n and m^* are related to the geometry of the lattice, giving MMs much greater flexibility than conventional materials.

For planar metamaterial design, an electrical variant of the SRR, termed Electrically coupled LC Resonator (ELC) as shown in Fig. 1.7a, has become the canonical structure for implementing negative values of permittivity (ϵ) for frequencies ranging from low RF to the optical [68, 79, 98–100, 104, 109–112, 151]. One of the fundamental reasons for using an electric structure such as the ELC is, so that the plane EM wave can be incident normal to the planar structure as shown in Fig. 1.7a with the electric field polarized perpendicular to the split gap. The modified symmetric geometry compared to conventional SRRs enables them to effectively cancel the magnetic moment and only couple to the incident electric field vector as shown in Fig. 1.7d–f [120, 121].

The electric response is shown in Fig. 1.7c similar to that of the magnetic response of the SRR in Fig. 1.6c, where ϵ_{eff} can be expressed as [100, 122, 123]

$$\epsilon_{\text{eff}}(\omega) = \epsilon_\infty - \frac{F\omega^2}{\omega^2 - \omega_0^2 - i\gamma\omega} \tag{1.5}$$

$$= \epsilon'_{\text{eff}} + i\epsilon''_{\text{eff}} \tag{1.6}$$

where ϵ_∞ is the frequency independent dielectric permittivity in the supporting dielectric substrate (e.g., FR4 or GaAs substrate). F is a geometrical factor, ω_0 is the resonance frequency, and γ is the resistive damping factor. As in the case of SRRs, if the Q of the resonator is high enough, the *real* part of Eq. (1.5) will result in a negative value around the resonant frequency, yielding an effective electric response from the ELC array.

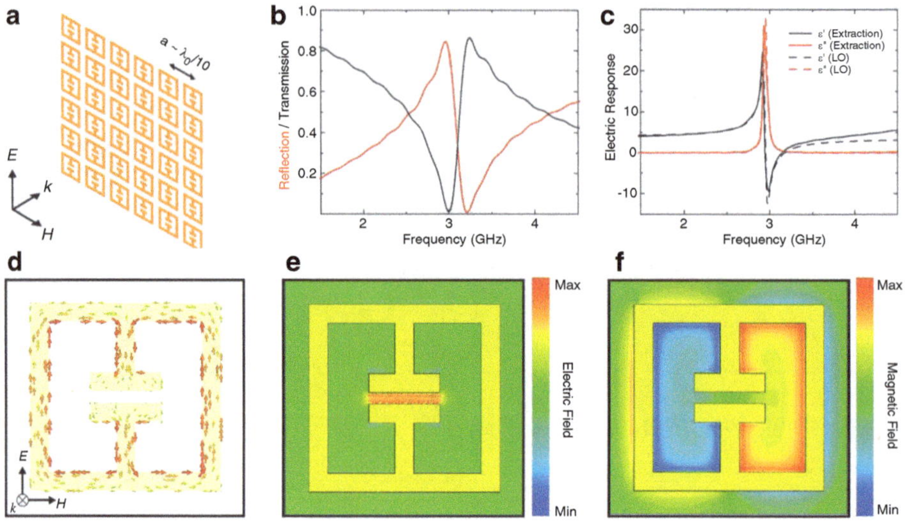

Fig. 1.7 Electrically coupled LC resonators (ELCs). (**a**) Incident electromagnetic wave propagates normal to the ELC array, where the element spacing a is 1/10 of the free space wavelength λ_0. (**b**) The simulated reflection and transmission for an ELC with $a = 10\,\text{mm}$. (**c**) The extracted electric response (*solid curves*) from simulation results shown in (**b**) and the response when fit with the theoretical model listed in Eq. (1.5), where $\epsilon_\infty = 4.3(\text{FR4})$, $F = 0.55$, $\omega_0 = 2\pi \times 2.96\,\text{GHz}$, and $\gamma = 0.05\,\text{s}^{-1}$. (**d**)–(**f**) Resonant field distributions in the plane of the metamaterial where (**d**) shows the resonant current density (**e**) the maximum electric field, and (**f**) plots maximum magnetic field [113]

1.2.3 Metamaterials: A Suitable Technology for Terahertz Devices

As explained in the previous section, metamaterials (MMs) are macroscopic composites of periodic structures whose electromagnetic (EM) response can be designed over a large portion of the EM spectrum by, to first order, simply scaling the dimensions of the periodic structures. MMs are essentially effective mediums requiring the periodic structures to be fraction of the wavelength ($\sim\lambda/10$) of the incident EM wave. For THz MMs, the unit cells are few tens of microns (3–300 μm) with critical feature sizes of few microns. They also require two critical materials for the construction, non-conducting substrate and metal layers separated from the substrate using dielectrics. These scales and the materials make them the ideal candidate for conventional microfabrication using a large pool of commercially available integrated circuit (IC) technologies, which allows for cost-effective, highly efficient means of production. The added benefit of using conventional microfabrication process; especially integrated circuit (IC) design processes, it allows researchers and engineers to select a vast range of solid-state devices to dynamically control the response as well as spectral selectivity of MM devices. Both of which are useful properties to develop novel THz imaging and communication devices previously not possible with available technologies.

In the last decade we have seen a rapid increase in metamaterial devices created for the THz regime using planar IC fabrication technologies for a wide range of applications including absorbers, quarter waveplates, switches/modulators, structurally reconfigurable MMs and MMs with memory effects.

1.2.3.1 Brief Overview of Metamaterial Based Terahertz Devices

The suitability of metamaterials as terahertz device has allowed researchers in the last decade to create MM devices for terahertz frequencies that are not easily obtainable in nature, most notably a magnetic response that is particularly rare at THz frequencies. One of the first THz MM was experimentally demonstrated by Yen et al. in 2004, showing a strong magnetic response at 1 THz, using a single planar double SRR array as shown in Fig. 1.8a [124]. Limitation to normal incident wave with planar magnetic SRRs is explained in Sect. 1.2.2.1.

Along with the fundamental mode of LC resonance in the SRRs or ELCs, higher mode dipole resonances associated with the sidebars of the SRR or ELC is present and may couple strongly with the fundamental mode depending on the design. The dipole resonance can be tuned by changing the distance between SRR sidebars, which are parallel to the electric field [125], as shown in Fig. 1.8b. Tuning the dipole resonance independent to the LC resonance is very important in applications where resonance reshaping due to LC-dipole coupling is undesirable.

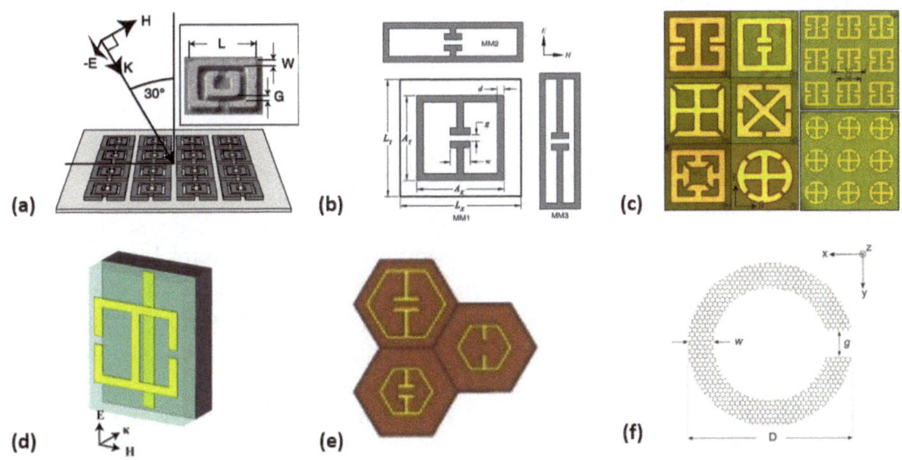

Fig. 1.8 Recent progress THz metamaterial devices. (**a**) Magnetically coupled SRRs [124]. (**b**) Tuned coupling between the LC and dipole resonances [125]. (**c**) THz electric MMs with symmetry geometries [121]. (**d**) Schematic of a THz MM absorber [79]. (**e**) Electric MMs with multiple resonances [127]. (**f**) Graphene based split-ring resonator [128]

Although SRRs can exhibit either purely negative electric or magnetic response for a chosen polarization of electric or magnetic field, the electric and magnetic resonances are coupled, resulting in a complex bianistropic EM response. A number of alternative SRR structures have been designed to suppress the magnetic response in favor of a pure electric response [121], as shown in Fig. 1.8c.

While most applications require the minimum loss related to the MMs, conversely, for many applications it would be desirable to maximize the loss. This is particularly important for THz frequencies where finding naturally occurring materials as an absorber for a desired frequency is difficult. MM-based absorbers have been demonstrated with a high absorptivity at terahertz frequencies [78, 79, 122, 126]. A single unit cell consists of an electrical ring resonator and a magnetic resonator as shown in Fig. 1.8d. The electric resonator is a standard ELC structure to cancel the magnetic moment and the magnetic resonator is realized by combining the center wire of the electric resonator on the top layer with a cut wire below it using a bottom layer metal. The EM responses are tuned to match the free-space impedance and minimize the transmission at a specific frequency.

One of the drawbacks of metamaterial devices is the narrow spectral range making it unsuitable for broadband applications. Efforts have been made to broaden the bandwidth by packing two or more resonators in a unit cell with different resonances [127], as shown in Fig. 1.8e.

The major limitation in developing high frequency metamaterial devices is the loss associated with the metal layer and the dielectric as well. Recently, studies have speculated the use of graphene to construct SRRs [128], as shown in Fig. 1.8f, allowing high confinement, long lifetimes and fast electrical tunability compared to the conventional gold structures. Recently, active circuits have been used to demonstrate loss compensation in microwave metamaterials [129, 130], but application of the same principle in terahertz frequencies is difficult and yet to be realized.

1.3 Overview of Terahertz Wave Modulators

Terahertz (THz) wave modulator is a key component in THz systems for wide range of applications from imaging to communication. A key to high-speed communication using THz wave is fast and efficient amplitude and/or phase modulator which is used to encode information in the carrier wave. It is also an important component in a single sensor THz imaging system where an array of THz wave modulators in the object plane is used to spatially encode the terahertz wave before being directed to the single sensor [131, 132]. A set of unique single sensor measurements corresponding to a set of coded apertures is used to reconstruct the image using a computational method known as compressed sensing (CS) [133]. A detailed discussion of such an imaging system is presented in Chap. 5.

Apart from amplitude and phase control, modulators can be used to control polarization state, spatial propagation direction, pulse shape, pulse length, and many more characteristic properties of electromagnetic waves. While active modulators are well established and considered standard devices in the optical realm, the THz frequency regime is still in great demand for highly efficient, fast, and versatile active wave modulators.

Modulators can be categorized by the physical quantity they control, as e.g. amplitude, phase, pulse length and shape, spectrum, spatial and temporal properties or by the technique or material system which is employed to modulate the wave. In most cases, modulators manipulate multiple properties of the THz wave at once, either on purpose or as a consequence of an undesired side effect. This section will briefly review some of the technologies developed till date, including metamaterials, for terahertz wave modulators.

In the past, semiconductors have been used for all-optical modulation of electromagnetic (EM) waves. Generally, laser pulses are incident on the semiconductor producing free carriers for time a period associated with the recombination time (τ). Typically, the plasma frequency (ω_p) of the carriers is above 1 THz ($\omega_p > 2\pi \times 1$ THz). Then, the surface looks "metallic" for frequencies below 1 THz acting like a reflective surface for time scales up to $t = \tau$. THz wave, co-incident on this area of high reflectivity, is thus modulated. This technique has been used since the 1990s to create THz modulators [134, 135]. In a more recent approach, spatially modulated laser light was used to induce one- and two-dimensional, tunable optical gratings in high-resistivity silicon to modulate terahertz waves [136, 137]. The same concept can be applied to metamaterials (MMs) where the resonance frequency of the metamaterial element is shifted by changing the capacitance of the split gap. The technique was used by Chen et al. to demonstrate a frequency-agile MM device, which is able to shift the center resonance frequency by 20% using external optical pumping [93], SEM photograph and measurement results shown in Fig. 1.9.

Although great progress in optically based THz modulators has been achieved, as described above, an all-electronic approach is an attractive proposition, especially with a view toward applications. It also allows for control circuit integration in the metamaterial device for achieving higher modulation speed [99]. Similar to the

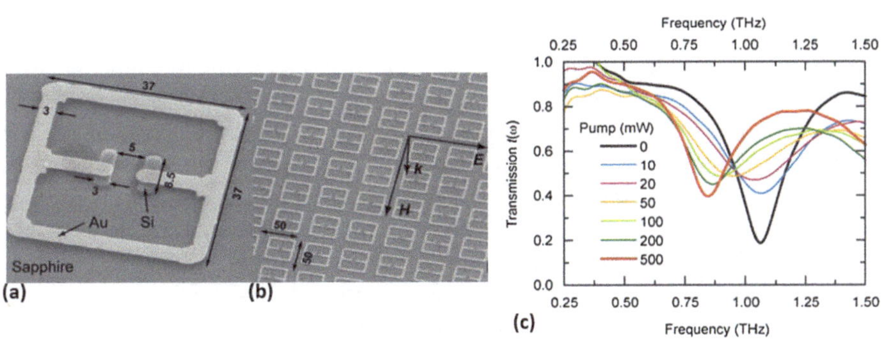

Fig. 1.9 Optically pumped frequency-agile THz MM. SEM images of (**a**) An individual unit cell and (**b**) a periodically patterned square array. (**c**) Experimental measurements as a function of photoexcitation power [93]

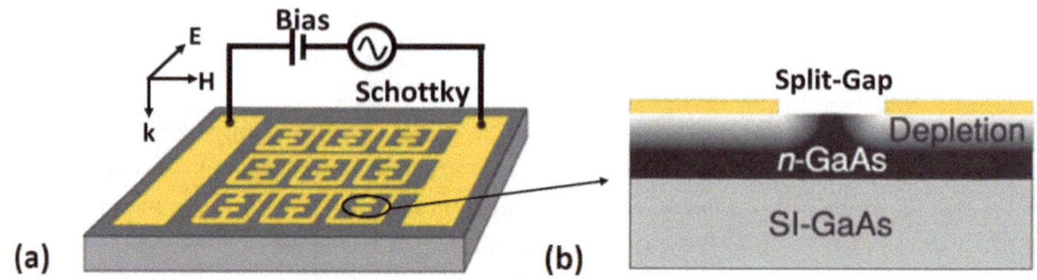

Fig. 1.10 Electrically controlled THz MM modulator. (**a**) A voltage bias applied between the Schottky and ohmic contacts controls the substrate charge carrier density near the split gaps, tuning the strength of the resonance. Orientation of the incident THz wave is indicated and the polarization of the electric field, E, magnetic field, H, and wave vector, k, are shown. (**b**) Diagram of the substrate and the depletion region near the split gap, where the *grey scale* indicates the free charge carrier density [151]

underlying principle of THz wave modulation by photo-doping of semiconductors, the carrier concentration in semiconductors can be changed by electric injection or depletion of charge carriers. Chen et al. first reported such a device [151]. A thin n-type GaAs layer was used to inject carrier into the split gap by applying a bias voltage across the metal layer and the n-type Gas layer, as shown in Fig. 1.10. Similar principle of shunting the split gap capacitance by optically pumping the substrate has also been previously reported [92, 94]. One of the limitations of such technique is slow modulation speed since an entire substrate is being modulated by a pulsed voltage source with a large intrinsic capacitor. A similar structure was fabricated in which a hole array functioned as the GaAs epilayer [138], with 2% change in transmission peak frequency. Another work fabricated a similar structure and demonstrated a maximum transmission modulation depth of 52% [139]. Ring apertures fabricated in silicon were shown to yield transmissive values of 60%, thus suggesting fabrication of THz modulators by CMOS architectures [140]. In the last decade the use of two-dimensional electron gases (2DEGs) in semiconductors has proven useful for control of THz waves and high electron mobility transistor (HEMT) has been a popular way to implement it [141].

More recently, it has been found that graphene is superior to semiconductors when used as an electrically driven modulator [142–146], due to its unique band structure which exhibits a linear dispersion relation between energy and crystal momentum. This results in extremely high carrier mobilities of up to $20{,}000\,\mathrm{cm^2V^{-1}s^{-1}}$ for both holes and electrons. Furthermore, outstandingly high carrier concentrations up to $1 \times 10^{14}\,\mathrm{cm^{-2}}$ can be obtained in graphene as well as extremely low carrier concentration at the Dirac point [146]. Thus, the carrier concentration can be tuned by applying an external electric field to the graphene layer. Sensale-Rodriguez et al. reported an electronically driven graphene terahertz wave modulator based on modification of the conductivity of graphene by controlling intraband transitions in a single graphene layer or alternately in graphene-semiconductor stacks [146]. The intensity modulation depth was reported to be about 16% in the frequency range from 570 to 630 GHz at 20 kHz modulation

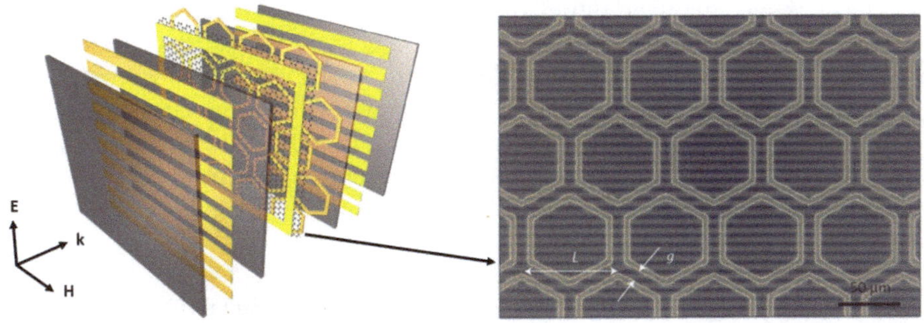

Fig. 1.11 Gate-controlled active graphene metamaterial. Schematic rendering of a gate-controlled active graphene metamaterial composed of an SLG deposited on a layer of hexagonal metallic meta-atoms. Metallic patterns and dielectric materials are represented by *yellow* and *grey*, respectively. The polarization of the incident terahertz wave is perpendicular to the line electrode, as indicated by the *arrows* [147]

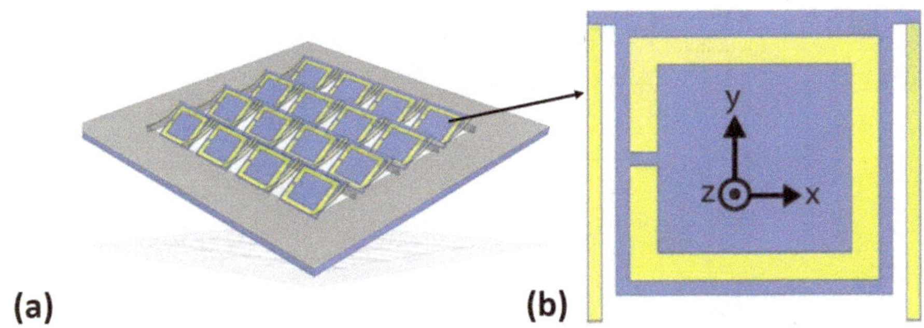

Fig. 1.12 Structurally reconfigurable THz MM fabricated on bimaterial cantilevers. (**a**) Schematic view of a portion of the metamaterial structure highlighting how the SRRs rotate as the cantilever legs bend. (**b**) Unit cell consisting of a split-ring resonator and cantilever legs [98]

speed. Lee et al. demonstrated terahertz wave switching in gate-controlled graphene metamaterials [147]. The graphene metamaterial was composed of single-layer graphene on top of a metamaterial with a hexagonal unit cell deposited on a polyimide substrate as depicted in Fig. 1.11. A maximum amplitude modulation depth of 90% was reported at the resonance frequency of 0.68 THz.

One method of THz modulation is to thermally tune the electrical conductivity and thus the optical response of semiconductors or metal oxides [148], special insulator materials with metallic phase transition [97], or superconductors [149]. An example of a thermally tunable metamaterial modulator is shown in Fig. 1.12 [98]. As a major disadvantage, thermal modulation is comparably slow with time constants in the range of several tens of milliseconds or longer.

Since most terahertz detectors today measure the incident power or the intensity of the THz wave, the terahertz modulators described so far are intensity modulators of the incident THz wave, either in transmission or reflection geometry.

But there is an important subset of THz imaging which benefits from phase-sensitive spectroscopic images for material identification. With progress in terahertz time-domain spectroscopy, quantum-cascade lasers, and continuous-wave terahertz spectrometers, phase information can now be extracted easily, even in compact THz systems making terahertz phase modulators a desirable THz component. Given the difficulty in building terahertz phase modulators, recent demonstration of a metamaterial solid-state terahertz phase modulator [96] is promising for building linearly controllable terahertz phase shifters.

Few more methods for designing terahertz wave modulators are compiled in a very comprehensive review by Rahm et al. in [150]. The results of all the methods are summarized very succinctly which is reproduced in Table 1.1.

Table 1.1 The table summarizes some characteristic quantities of electrically, optically, and thermally driven terahertz wave modulators for various material systems

Method	Material	τ/f_{Mod}	Frequency	M_{A2}	Ref.
Electrical	GaAs/AlGaAs(2DEG)	–	0.1–2 THz	3%	[141]
Electrical	Meta/nGaAs	–	0.72 THz	30%	[151]
Electrical	GaAs (hole array)	–	–	33%	[139]
Electrical	Meta/nGaAs	100 kHz	0.63 THz	36%	[152]
Electrical	Meta/nGaAs	2 MHz	–	–	[95]
Electrical	Meta/HEMT	10 MHz	0.46 THz	33%	[99]
Electrical	Graphene	–	0.57–0.63 THz	9%	[146]
Electrical	Meta/Graphene	–	0.68 THz	59%	[147]
Optical	Silicon	28 ms	2.5 THz	–	[153]
Optical	Silicon	430 ns	2.5 THz	–	[134]
Optical	GaAs	10 ns	1.4 THz	–	[154]
Optical	Silicon	5 ns	1.4 THz	–	[135]
Optical	InSb (grating)	ps	–	0.25%,1%	[155]
Optical	Meta/GaAs	–	0.56 THz	80%	[92]
Optical	Meta/ErAs/GaAs	20 ps	0.75 THz	35%	[94]
Optical	Meta/SOS	–	0.6 THz	70%	[156]
Thermal	Meta/VO2	–	1.0 THz	70%	[157]
Thermal	Meta/YBCO	–	0.61–0.55 THz	90%	[149]

The modulators are evaluated with respect to the modulation bandwidth f_{Mod} and the relaxation time τ as well as the operating frequency f. Furthermore, approximate values of the modulation depth are estimated. The modulation depth was defined by $M_{A2} = (E_{\max} - E_{\min})/E_{\max}$, where $E_{\max}$ is the maximal transmitted electric field amplitude and $E_{\min}$ is the minimal transmitted field amplitude. The last column refers to the reference number as listed in the reference section [150]

References

1. P. Siegel, IEEE Trans. Microwave Theory Tech. **50**(3), 910 (2002)
2. G.P. Williams, Rep. Prog. Phys. **69**(2), 301 (2006)
3. H. Tao, W. Padilla, X. Zhang, R. Averitt, IEEE J. Selected Topics Quantum Electron. **17**(1), 92 (2011)
4. M. Tonouchi, Nat. Photonics **1**(2), 97 (2007)
5. J.M. Chamberlain, Philos. Trans. R. Soc. Lond. A Math. Phys. Eng. Sci. **362**(1815), 199 (2004)
6. A. Maestrini, J. Ward, J. Gill, H. Javadi, E. Schlecht, G. Chattopadhyay, F. Maiwald, N.R. Erickson, I. Mehdi, IEEE Microw. Wirel. Compon. Lett. **14**(6), 253 (2004)
7. D. Mittleman, *Sensing with Terahertz Radiation* (Springer, Berlin, 2013)
8. W.S. Holland, J.S. Greaves, B. Zuckerman, R.A. Webb, C. McCarthy, I.M. Coulson, D.M. Walther, W.R.F. Dent, W.K. Gear, I. Robson, Nature **392**(6678), 788 (1998)
9. N.C. Luhmann Jr., W.A. Peebles, Rev. Sci. Instrum. **55**(3), 279 (1984)
10. M.V. Exter, C. Fattinger, D. Grischkowsky, Optics Lett. **14**(20), 1128 (1989)
11. M. van Exter, D.R. Grischkowsky, IEEE Trans. Microwave Theory Tech. **38**(11), 1684 (1990)
12. R.A. Kaindl, M.A. Carnahan, J. Orenstein, D.S. Chemla, H.M. Christen, H.Y. Zhai, M. Paranthaman, D.H. Lowndes, Phys. Rev. Lett. **88**(2), 027003 (2001)
13. A.G. Markelz, A. Roitberg, E.J. Heilweil, Chem. Phys. Lett. **320**(12), 42 (2000)
14. S.P. Mickan, A. Menikh, H. Liu, C.A. Mannella, R. MacColl, D. Abbott, J. Munch, X.C. Zhang, Phys. Med. Biol. **47**(21), 3789 (2002)
15. D.M. Mittleman, M. Gupta, R. Neelamani, R.G. Baraniuk, J.V. Rudd, M. Koch, Appl. Phys. B **68**(6), 1085 (1999)
16. W.L. Chan, J. Deibel, D.M. Mittleman, Rep. Prog. Phys. **70**(8), 1325 (2007)
17. N. Karpowicz, H. Zhong, C. Zhang, K.I. Lin, J.S. Hwang, J. Xu, X.C. Zhang, Appl. Phys. Lett. **86**(5), 054105 (2005)
18. K. Kawase, Y. Ogawa, Y. Watanabe, H. Inoue, Opt. Express **11**(20), 2549 (2003)
19. T.M. Korter, D.F. Plusquellic, Chem. Phys. Lett. **385**(12), 45 (2004)
20. R.M. Woodward, V.P. Wallace, R.J. Pye, B.E. Cole, D.D. Arnone, E.H. Linfield, M. Pepper, J. Invest. Dermatol. **120**(1), 72 (2003)
21. B. Ferguson, X.C. Zhang, Nat. Mater. **1**(1), 26 (2002)
22. P.Y. Han, G.C. Cho, X.C. Zhang, Opt. Lett. **25**(4), 242 (2000)
23. T. Kleine-Ostmann, T. Nagatsuma, J. Infrared Millimeter Terahertz Waves **32**(2), 143 (2011)
24. R. Piesiewicz, T. Kleine-Ostmann, N. Krumbholz, D. Mittleman, M. Koch, J. Schoebel, T. Kurner, IEEE Antennas Propag. Mag, **49**(6), 24 (2007)
25. J. Federici, L. Moeller, J. Appl. Phys. **107**(11), 111101 (2010)
26. T. Nagatsuma, H.J. Song, Y. Fujimoto, K. Miyake, A. Hirata, K. Ajito, A. Wakatsuki, T. Furuta, N. Kukutsu, Y. Kado, in *Technical Digest of IEEE International Topical Meeting on Microwave Photonics* (2009)
27. H.A. Gebbie, N.W.B. Stone, F.D. Findlay, Nature **202**, 685 (1964)
28. T.S. Hartwick, D.T. Hodges, D.H. Barker, F.B. Foote, Appl. Opt. **15**(8), 1919 (1976)
29. D.H. Auston, A.M. Glass, P. LeFur, Appl. Phys. Lett. **23**(1), 47 (1973)
30. D.H. Auston, K.P. Cheung, J. Opt. Soc. Am. B **2**(4), 606 (1985)
31. A.P. DeFonzo, M. Jarwala, C. Lutz, Appl. Phys. Lett. **50**(17), 1155 (1987)
32. P.R. Smith, D.H. Auston, M.C. Nuss, IEEE J. Quantum Electron. **24**(2), 255 (1988)
33. C. Fattinger, D. Grischkowsky, Appl. Phys. Lett. **54**(6), 490 (1989)
34. D. Grischkowsky, S. Keiding, M.V. Exter, C. Fattinger, J. Opt. Soc. Am. B **7**(10), 2006 (1990)
35. B.B. Hu, M.C. Nuss, Opt. Lett. **20**(16), 1716 (1995)
36. D.M. Mittleman, S. Hunsche, L. Boivin, M.C. Nuss, Opt. Lett. **22**(12), 904 (1997)
37. K. McClatchey, M.T. Reiten, R.A. Cheville, Appl. Phys. Lett. **79**(27), 4485 (2001)
38. T.D. Dorney, J.L. Johnson, J.V. Rudd, R.G. Baraniuk, W.W. Symes, D.M. Mittleman, Opt. Lett. **26**(19), 1513 (2001)

39. B. Ferguson, S. Wang, D. Gray, D. Abbot, X.C. Zhang, Opt. Lett. **27**(15), 1312 (2002)
40. A.W. Lee, Q. Hu, Opt. Lett. **30**(19), 2563 (2005)
41. Q. Wu, T.D. Hewitt, X.C. Zhang, Appl. Phys. Lett. **69**(8), 1026 (1996)
42. M. Helm, E. Colas, P. England, F. DeRosa, S.J. Allen Jr., Appl. Phys. Lett. **53**(18), 1714 (1988)
43. J. Faist, F. Capasso, D.L. Sivco, C. Sirtori, A.L. Hutchinson, A.Y. Cho, Science **264**(5158), 553 (1994)
44. R. Khler, A. Tredicucci, F. Beltram, H.E. Beere, E.H. Linfield, A.G. Davies, D.A. Ritchie, R.C. Iotti, F. Rossi, Nature **417**(6885), 156 (2002)
45. A.W.M. Lee, Q. Qin, S. Kumar, B.S. Williams, Q. Hu, J.L. Reno, Appl. Phys. Lett. **89**(14), 141125 (2006)
46. S. Fathololoumi, E. Dupont, C.W.I. Chan, Z.R. Wasilewski, S.R. Laframboise, D. Ban, A. Mtys, C. Jirauschek, Q. Hu, H.C. Liu, Opt. Express **20**(4), 3866 (2012)
47. L. Li, L. Chen, J. Zhu, J. Freeman, P. Dean, A. Valavanis, A.G. Davies, E.H. Linfield, Electron. Lett. **50**(4), 309 (2014)
48. T. Phillips, J. Keene, Proc. IEEE **80**(11), 1662 (1992)
49. J. Darmo, V. Tamosiunas, G. Fasching, J. Krll, K. Unterrainer, M. Beck, M. Giovannini, J. Faist, C. Kremser, P. Debbage, Opt. Express **12**(9), 1879 (2004)
50. S.M. Kim, F. Hatami, J.S. Harris, A.W. Kurian, J. Ford, D. King, G. Scalari, M. Giovannini, N. Hoyler, J. Faist, G. Harris, Appl. Phys. Lett. **88**(15), 153903 (2006)
51. S. Barbieri, J. Alton, C. Baker, T. Lo, H.E. Beere, D. Ritchie, Opt. Express **13**(17), 6497 (2005)
52. K.L. Nguyen, M.L. Johns, L.F. Gladden, C.H. Worrall, P. Alexander, H.E. Beere, M. Pepper, D.A. Ritchie, J. Alton, S. Barbieri, E.H. Linfield, Opt. Express **14**(6), 2123 (2006)
53. H.A. Gebbie, in *Proceedings Symposium on Submillimeter Waves* (Polytechnic Press of the Polytechnic Institute of Brooklyn, Brooklyn, 1970)
54. A. Hirata, T. Kosugi, H. Takahashi, R. Yamaguchi, F. Nakajima, T. Furuta, H. Ito, H. Sugahara, Y. Sato, T. Nagatsuma, IEEE Trans. Microwave Theory Techn. **54**(5), 1937 (2006)
55. A. Hirata, H. Takahashi, R. Yamaguchi, T. Kosugi, K. Murata, T. Nagatsuma, N. Kukutsu, Y. Kado, J. Lightwave Technol. **26**(15), 2338 (2008)
56. T. Kosugi et al., 120-GHz Tx/Rx waveguide modules for 10-Gbit/s wireless link system, in *The Compound Semiconductor Integrated Circuit Symposium (CSIC)* (IEEE, 2006), pp. 25–28
57. N. Engheta, R.W. Ziolkowski, *Metamaterials: Physics and Engineering Explorations* (Wiley, New York, 2006)
58. J.C. Bose, Proc. R. Soc. Lond. **63**, 146 (1898)
59. I.V. Lindell, A.H. Sihvola, J. Kurkijarvi, IEEE Antennas Propag. Mag. **34**(3), 24 (1992)
60. L. Mandlshtam, Zhurnal Eksperimentalnoii Teoreticheskoi fiziki **15**, 476 (1945)
61. W.E. Kock, Bell Syst. Tech. J. **27**(1), 58 (1948)
62. G.D. Malyuzhinets, Zhurnal technicheskoi fiziki **21**(8), 940 (1951)
63. D.V. Sivukhin, Opt. Spektrosk **3**(308–331), 2 (1957)
64. S.A. Tretyakov, in *Negative Refraction: Revisiting Electromagnetics from Microwave to Optics, EPFL Latsis Symposium, Lausanne* (2005), pp. 30–35
65. V. Veselago, Soviet Phys. Uspekh **10**(4), 509 (1968)
66. J.B. Pendry, A.J. Holden, W.J. Stewart, I. Youngs, Phys. Rev. Lett. **76**(25), 4773 (1996)
67. D. Smith, D. Vier, W. Padilla, S. Nemat-Nasser, S. Schultz, Appl. Phys. Lett. **75**(10), 1425 (1999)
68. D.R. Smith, W.J. Padilla, D.C. Vier, S.C. Nemat-Nasser, S. Schultz, Phys. Rev. Lett. **84**(18), 4184 (2000)
69. D.R. Smith, J.B. Pendry, M.C.K. Wiltshire, Science **305**(5685), 788 (2004)
70. J. Pendry, Nature **460**(7255), 579 (2009)
71. J. Pendry, Nat. Mater. **5**(8), 599 (2006)
72. W.J. Padilla, D.N. Basov, D.R. Smith, Mater. Today **9**(7–8), 28 (2006)
73. V.M. Shalaev, Nat. Photonics **1**(1), 41 (2007)

74. K. Aydin, I. Bulu, E. Ozbay, Opt. Express **13**(22), 8753 (2005)
75. J.B. Pendry, Phys. Rev. Lett. **85**(18), 3966 (2000)
76. N. Fang, H. Lee, C. Sun, X. Zhang, Science **308**(5721), 534 (2005)
77. X. Zhang, Z. Liu, Nat. Mat. **7**(6), 435 (2008)
78. N.I. Landy, S. Sajuyigbe, J.J. Mock, D.R. Smith, W.J. Padilla, Phys. Rev. Lett. **100**(20), 207402 (2008)
79. H. Tao, N.I. Landy, C.M. Bingham, X. Zhang, R.D. Averitt, W.J. Padilla, Opt. Express **16**(10), 7181 (2008)
80. J.B. Pendry, D. Schurig, D.R. Smith, Science **312**(5781), 1780 (2006)
81. B. Kant, D. Germain, A. de Lustrac, Phys. Rev. B **80**(20), 201104 (2009)
82. H. Tao, N. Landy, K. Fan, A. Strikwerda, W. Padilla, R. Averitt, X. Zhang, in *Electron Devices Meeting, 2008. IEDM 2008. IEEE International* (2008), pp. 1–4.
83. D. Schurig, J. Mock, B. Justice, S. Cummer, J. Pendry, A. Starr, D. Smith, Science **314**(5801), 977 (2006)
84. X.Q. Lin, T.J. Cui, J.Y. Chin, X.M. Yang, Q. Cheng, R. Liu, Appl. Phys. Lett. **92**(13), 131904 (2008)
85. R. Liu, X.M. Yang, J.G. Gollub, J.J. Mock, T.J. Cui, D.R. Smith, Appl. Phys. Lett. **94**(7), 073506 (2009)
86. D.R. Smith, J.J. Mock, A.F. Starr, D. Schurig, Phys. Rev. E **71**(3), 036609 (2005)
87. R. Liu, C. Ji, J.J. Mock, T. Cui, D. Smith, in *2008 International Workshop on Metamaterials* (2008), pp. 248–250
88. A.K. Iyer, P.C. Kremer, G.V. Eleftheriades, Opt. Express **11**(7), 696 (2003)
89. A.A. Oliner, in *IEEE-AP-S Digest* (San Anotnio, TX, 2002), p. 41
90. C. Caloz, T. Itoh, IEEE **2**, 412–415 (2002)
91. C. Caloz, T. Itoh, *Electromagnetic Metamaterials: Transmission Line Theory and Microwave Applications: The Engineering Approach* (Wiley, New York, 2006)
92. W.J. Padilla, A.J. Taylor, C. Highstrete, M. Lee, R.D. Averitt, Phys. Rev. Lett. **96**(10), 107401 (2006)
93. H.T. Chen, J.F. O'Hara, A.K. Azad, A.J. Taylor, R.D. Averitt, D.B. Shrekenhamer, W.J. Padilla, Nat. Photon **2**(5), 295 (2008)
94. H.T. Chen, W.J. Padilla, J.M.O. Zide, S.R. Bank, A.C. Gossard, A.J. Taylor, R.D. Averitt, Opt. Lett. **32**(12), 1620 (2007)
95. H.T. Chen, S. Palit, T. Tyler, C.M. Bingham, J.M.O. Zide, J.F. OHara, D.R. Smith, A.C. Gossard, R.D. Averitt, W.J. Padilla, N.M. Jokerst, A.J. Taylor, Appl. Phys. Lett. **93**(9), 091117 (2008)
96. H.T. Chen, W.J. Padilla, M.J. Cich, A.K. Azad, R.D. Averitt, A.J. Taylor, Nat. Photonics **3**(3), 148 (2009)
97. T. Driscoll, S. Palit, M.M. Qazilbash, M. Brehm, F. Keilmann, B.G. Chae, S.J. Yun, H.T. Kim, S.Y. Cho, N.M. Jokerst, D.R. Smith, D.N. Basov, Appl. Phys. Lett. **93**(2), 024101 (2008)
98. H. Tao, A.C. Strikwerda, K. Fan, W.J. Padilla, X. Zhang, R.D. Averitt, Phys. Rev. Lett. **103**(14), 147401 (2009)
99. D. Shrekenhamer, S. Rout, A.C. Strikwerda, C. Bingham, R.D. Averitt, S. Sonkusale, W.J. Padilla, Opt. Express **19**(10), 9968 (2011)
100. R. Liu, T.J. Cui, D. Huang, B. Zhao, D.R. Smith, Phys. Rev. E **76**(2), 026606 (2007)
101. D.R. Smith, J.B. Pendry, J. Opt. Soc. Am. B **23**(3), 391 (2006)
102. A. Demetriadou, J.B. Pendry, J. Phys. Condens. Matter **20**(29), 295222 (2008)
103. M.C.K. Wiltshire, J.B. Pendry, J.V. Hajnal, J. Phys. Condens. Matter **21**(29), 292201 (2009)
104. J. Pendry, A. Holden, D. Robbins, W. Stewart, IEEE Trans. Microwave Theory Tech. **47**(11), 2075 (1999)
105. G. Dolling, C. Enkrich, M. Wegener, J.F. Zhou, C.M. Soukoulis, S. Linden, Opt. Lett. **30**(23), 3198 (2005)
106. V.M. Shalaev, W. Cai, U.K. Chettiar, H.K. Yuan, A.K. Sarychev, V.P. Drachev, A.V. Kildishev, Opt. Lett. **30**(24), 3356 (2005)
107. C. Rockstuhl, C. Menzel, T. Paul, T. Pertsch, F. Lederer, Phys. Rev. B **78**(15), 155102 (2008)

108. J. Valentine, S. Zhang, T. Zentgraf, E. Ulin-Avila, D.A. Genov, G. Bartal, X. Zhang, Nature **455**(7211), 376 (2008)
109. D. Shrekenhamer, W. Xu, S. Venkatesh, D. Schurig, S. Sonkusale, W.J. Padilla, Phys. Rev. Lett. **109**(17), 177401 (2012)
110. D. Li, Y. Xie, J. Zhang, J. Li, Z. Chen, J. Electromagn. Waves Appl. **22**(10), 1420 (2008)
111. C. Rockstuhl, T. Zentgraf, H. Guo, N. Liu, C. Etrich, I. Loa, K. Syassen, J. Kuhl, F. Lederer, H. Giessen, Appl. Phys. B **84**(1–2), 219 (2006)
112. J. Wang, S. Qu, Z. Xu, J. Zhang, H. Ma, Y. Yang, C. Gu, Photonics Nanostruct. Fundam. Appl. **7**(2), 108 (2009)
113. D. Shrekenhamer, *Dynamic Control of Metamaterials at Terahertz Frequencies*, Ph.D., Boston College, 2013
114. S. OBrien, J.B. Pendry, J. Phys. Condens. Matter **14**(25), 6383 (2002)
115. R. Bracewell, Wirel. Eng. **31**, 320 (1954)
116. W. Rotman, IRE Trans. Antennas Propag. **10**(1), 82 (1962)
117. R. Ulrich, Infrared Phys. **7**(1), 37 (1967)
118. T. Timusk, P. Richards, Appl. Opt. **20**(8), 1355 (1981)
119. J. Pendry, A. Holden, D. Robbins, W. Stewart, J. Phys. Condens. Matter **10**(22), 4785 (1998)
120. D. Schurig, J. Mock, D.R. Smith, Appl. Phys. Lett. **88**(4), 041109 (2006)
121. W.J. Padilla, M.T. Aronsson, C. Highstrete, M. Lee, A.J. Taylor, R.D. Averitt, Phys. Rev. B **75**(4), 041102 (2007)
122. N.I. Landy, C.M. Bingham, T. Tyler, N. Jokerst, D.R. Smith, W.J. Padilla, Phys. Rev. B **79**(12), 125104 (2009)
123. B.J. Arritt, D.R. Smith, T. Khraishi, J. Appl. Phys. **109**(7), 073512 (2011)
124. T.J. Yen, W.J. Padilla, N. Fang, D.C. Vier, D.R. Smith, J.B. Pendry, D.N. Basov, X. Zhang, Science (New York, NY) **303**(5663), 1494 (2004)
125. A.K. Azad, A.J. Taylor, E. Smirnova, J.F. OHara, Appl. Phys. Lett. **92**(1), 011119 (2008)
126. C.M. Watts, X. Liu, W.J. Padilla, Adv. Mater. **24**(23), OP98 (2012)
127. C.M. Bingham, H. Tao, X. Liu, R.D. Averitt, X. Zhang, W.J. Padilla, Opt. Express **16**(23), 18565 (2008)
128. N. Papasimakis, S. Thongrattanasiri, N.I. Zheludev, F.G.d. Abajo, Light Sci. Appl. **2**(7), e78 (2013)
129. W. Xu, W.J. Padilla, S. Sonkusale, arXiv:1204.6067 [physics] (2012)
130. D. Ye, K. Chang, L. Ran, H. Xin, Nat. Commun. **5** (2014)
131. C.M. Watts, D. Shrekenhamer, J. Montoya, G. Lipworth, J. Hunt, T. Sleasman, S. Krishna, D.R. Smith, W.J. Padilla, Nat. Photonics **8**, 605 (2014)
132. D. Shrekenhamer, C.M. Watts, W.J. Padilla, Opt. Express **21**(10), 12507 (2013)
133. D. Donoho, IEEE Trans. Inf. Theory **52**(4), 1289 (2006)
134. T. Vogel, G. Dodel, E. Holzhauer, H. Salzmann, A. Theurer, Appl. Opt. **31**(3), 329 (1992)
135. T. Nozokido, H. Minamide, K. Mizuno, Electron. Commun. Jpn. (Part II: Electron.) **80**(6), 1 (1997)
136. T. Okada, K. Tanaka, Sci. Rep. **1**, 121 (2011)
137. S. Busch, B. Scherger, M. Scheller, M. Koch, Opt. Lett. **37**(8), 1391 (2012)
138. E.A. Shaner, J.G. Cederberg, D. Wasserman, Appl. Phys. Lett. **91**(18), 181110 (2007)
139. H.T. Chen, H. Lu, A.K. Azad, R.D. Averitt, A.C. Gossard, S.A. Trugman, J.F. OHara, A.J. Taylor, Opt. Express **16**(11), 7641 (2008)
140. J. Shu, C. Qiu, V. Astley, D. Nickel, D.M. Mittleman, Q. Xu, Opt. Express **19**(27), 26666 (2011)
141. T. Kleine-Ostmann, P. Dawson, K. Pierz, M. Koch, Appl. Phys. Lett. **84**(18), 3555 (2004)
142. L. Ju, B. Geng, J. Horng, C. Girit, M. Martin, Z. Hao, H.A. Bechtel, X. Liang, A. Zettl, Y.R. Shen, F. Wang, Nat. Nanotechnol. **6**(10), 630 (2011)
143. B. Sensale-Rodriguez, T. Fang, R. Yan, M.M. Kelly, D. Jena, L. Liu, H.G. Xing, Appl. Phys. Lett. **99**(11), 113104 (2011)
144. C.C. Lee, S. Suzuki, W. Xie, T.R. Schibli, Opt. Express **20**(5), 5264 (2012)

145. M. Liu, X. Yin, E. Ulin-Avila, B. Geng, T. Zentgraf, L. Ju, F. Wang, X. Zhang, Nature **474**(7349), 64 (2011)
146. B. Sensale-Rodriguez, R. Yan, M.M. Kelly, T. Fang, K. Tahy, W.S. Hwang, D. Jena, L. Liu, H.G. Xing, Nat. Commun. **3**, 780 (2012)
147. S.H. Lee, M. Choi, T.T. Kim, S. Lee, M. Liu, X. Yin, H.K. Choi, S.S. Lee, C.G. Choi, S.Y. Choi, X. Zhang, B. Min, Nat. Mater. **11**(11), 936 (2012)
148. J.G. Rivas, M. Kuttge, H. Kurz, P.H. Bolivar, J.A. Snchez-Gil, Appl. Phys. Lett. **88**(8), 082106 (2006)
149. H.T. Chen, H. Yang, R. Singh, J.F. OHara, A.K. Azad, S.A. Trugman, Q.X. Jia, A.J. Taylor, Phys. Rev. Lett. **105**(24), 247402 (2010)
150. M. Rahm, J.S. Li, W.J. Padilla, J. Infrared Millimeter Terahertz Waves **34**(1), 1 (2013)
151. H.T. Chen, W.J. Padilla, J.M.O. Zide, A.C. Gossard, A.J. Taylor, R.D. Averitt, Nature **444**(7119), 597 (2006)
152. O. Paul, C. Imhof, B. Lgel, S. Wolff, J. Heinrich, S. Hfling, A. Forchel, R. Zengerle, R. Beigang, M. Rahm, Opt. Express **17**(2), 819 (2009)
153. H. Alius, G. Dodel, Infrared Phys. **32**, 1 (1991)
154. T. Nozokido, H. Minamide, K. Mizuno, RIKEN Rev. **11**(11), 11–12 (1995)
155. E. Hendry, M.J. Lockyear, J.G. Rivas, L. Kuipers, M. Bonn, Phys. Rev. B **75**(23), 235305 (2007)
156. D.R. Chowdhury, R. Singh, J.F. OHara, H.T. Chen, A.J. Taylor, A.K. Azad, Appl. Phys. Lett. **99**(23), 231101 (2011)
157. T. Driscoll, H.T. Kim, B.G. Chae, B.J. Kim, Y.W. Lee, N.M. Jokerst, S. Palit, D.R. Smith, M.D. Ventra, D.N. Basov, Science **325**(5947), 1518 (2009)

Chapter 2
Background Theory

This chapter reviews some of the fundamental electromagnetic principles for a basic understanding of metamaterials and metamaterials as terahertz modulator. Section 2.1 covers the basic electromagnetic properties of materials with non-positive dielectric parameters, permitivity (ϵ) and permeability (μ). In Sect. 2.2, basic Lorentz oscillator model for permitivity is developed to illustrate the anomalous dispersion behavior that is fundamental to the modulator design. Finally, the basic principle of wave modulation using metamaterials is formulated in Sect. 2.3.

Complete and rigorous electromagnetic analysis is beyond the scope of this book. For such in-depth analysis there is an excellent collection of standard textbooks such as Landau et al. [1], Jackson [2], Kong [3]. For detailed analysis on optical properties of materials, an excellent reference is Wooten [4]. For further exploration of metamaterials, readers are referred to texts on metamaterials such as Ramakrishna et al. [5], Tretyakov [6], Eleftheriades et al. [7], Caloz et al. [8], Engheta et al. [9], Pendry [10].

2.1 Plane Waves in a Nonconducting Medium

Material response to electromagnetic waves can fundamentally be expressed by the Maxwell equations at the atomic length scales. However, in most cases we are not interested in the fast variation of electric and magnetic fields at those microscopic scales, instead a macroscopic description is sufficiently accurate. At a macroscopic level, the Maxwell equations are written as

$$\nabla \cdot \mathbf{D} = \rho \tag{2.1}$$

$$\nabla \cdot \mathbf{B} = 0 \tag{2.2}$$

$$\nabla \times \mathbf{E} + \frac{\partial \mathbf{B}}{\partial t} = 0 \tag{2.3}$$

S. Rout, S. Sonkusale, *Active Metamaterials*, DOI 10.1007/978-3-319-52219-7_2

$$\nabla \times \mathbf{H} - \frac{\partial \mathbf{D}}{\partial t} = \mathbf{J} \tag{2.4}$$

where **E** and **H** are the macroscopic electric and magnetic fields, **D** is the displacement field and **B** is the magnetic induction. Similarly, **J** and **j** are the macroscopic net charge and current densities. These macroscopic expressions are averaged over lengths that is large compared to inter-molecular distance but still only a fraction of wavelength of the applied fields, resulting in a homogeneous approximation of the material. A detailed derivation of these equations from microscopic Maxwell equations can be found in Jackson [2].

The metamaterials discussed in this text can be macroscopically approximated as dielectrics or nonconductors that are a devoid of free charge, and the Maxwell equations (2.1)–(2.4) can be reduced to

$$\nabla \cdot \mathbf{D} = 0 \tag{2.5}$$

$$\nabla \cdot \mathbf{B} = 0 \tag{2.6}$$

$$\nabla \times \mathbf{E} + \frac{\partial \mathbf{B}}{\partial t} = 0 \tag{2.7}$$

$$\nabla \times \mathbf{H} - \frac{\partial \mathbf{D}}{\partial t} = 0 \tag{2.8}$$

In most materials, which are linear and isotropic, the displacement field **D** is directly proportional to the applied electric field **E**:

$$\mathbf{D} = \epsilon \mathbf{E} \tag{2.9}$$

where ϵ which is dispersive, i.e., function of the frequency ω, $\mathbf{D} = \epsilon(\omega)\mathbf{E}$. The dispersion is associated with the inertia of the dipoles due to the mass of electrons. In addition to being dispersive, ϵ also requires to be a complex function on the account of causality where the imaginary part is attributed to the loss or the absorbed energy in the medium. For a succinct proof of ϵ as a dispersive and complex function, the reader is referred to Ramamkrishna et al. [5].

A similar analysis holds true for the magnetic permeability,

$$\mathbf{B} = \mu \mathbf{H} \tag{2.10}$$

where is μ is dispersive and a complex function as well.

The causality and analyticity domain of $\epsilon(\omega)/\epsilon_0$ allows the use of Cauchy's theorem to relate the real and the imaginary part of $\epsilon(\omega)/\epsilon_0$, known as the Kramers–Kronig relations, and is expressed as (Jackson [2])

$$\mathrm{Re}[\epsilon(\omega)/\epsilon_0] = 1 + \frac{1}{\pi} \mathrm{P} \int_{-\infty}^{\infty} \frac{\mathrm{Im}[\epsilon(\omega')/\epsilon_0]}{\omega' - \omega} d\omega' \tag{2.11}$$

$$\mathrm{Im}[\epsilon(\omega)/\epsilon_0] = -\frac{1}{\pi}\,\mathrm{P}\int_{-\infty}^{\infty}\frac{\mathrm{Re}[\epsilon(\omega')/\epsilon_0]-1}{\omega'-\omega}d\omega' \tag{2.12}$$

where P means the principal value of the Cauchy's integral. This relation is regarded as very fundamental to the dispersion nature of ϵ. This relation allows the calculation of Re[$\epsilon(\omega)$] from the experimental data of Im[$\epsilon(\omega)$] from absorption experiments.

For sinusoidal solutions ($e^{-i\omega t}$), the equations for the $\mathbf{E}(\omega, \mathbf{x})$,etc.

$$\begin{aligned} \nabla\cdot\mathbf{B} &= 0, \qquad \nabla\times\mathbf{E} - i\omega\mathbf{B} = 0 \\ \nabla\cdot\mathbf{D} &= 0, \qquad \nabla\times\mathbf{H} + i\omega\mathbf{D} = 0 \end{aligned} \tag{2.13}$$

For uniform isotropic linear media, $\mathbf{D} = \epsilon\mathbf{E}, \mathbf{B} = \mu\mathbf{H}$ where ϵ and μ in general maybe complex functions of ω. For real and positive ϵ and μ (no losses), **D** and **B** can be substituted in Eq. (2.13) to get the Helmholtz wave equation (See Appendix A.1)

$$\left(\nabla^2 + \mu\epsilon\omega^2\right)\begin{Bmatrix}\mathbf{E}\\ \mathbf{B}\end{Bmatrix} = 0 \tag{2.14}$$

A plane-wave solution, traveling in the x direction, that satisfies both the Maxwell's equation (2.13) and the Helmholtz's equation (2.14) can be shown to be

$$\begin{aligned} \mathbf{E}(\mathbf{x}, t) &= \mathbf{E}_0 e^{i(kx-\omega t)} \\ \mathbf{B}(\mathbf{x}, t) &= \mathbf{B}_0 e^{i(kx-\omega t)} \end{aligned} \tag{2.15}$$

where k is the wave number

$$k = \sqrt{\epsilon\mu}\,\omega \tag{2.16}$$

The *phase velocity* of the wave is

$$v = \frac{\omega}{k} = \frac{1}{\sqrt{\epsilon\mu}} = \frac{c}{n} \tag{2.17}$$

The quantity n is called the *index of refraction* that can be expressed as

$$n = \sqrt{\frac{\mu}{\mu_0}\frac{\epsilon}{\epsilon_0}} = \sqrt{\mu_r\epsilon_r} \tag{2.18}$$

The wave impedance Z can be expressed as

$$Z = \frac{E_0}{H_0} = \frac{k}{\omega\epsilon} = \frac{1}{v\epsilon} = \sqrt{\frac{\mu}{\epsilon}} = \zeta Z_0 \quad \zeta = \frac{\mu_r}{\epsilon_r} \tag{2.19}$$

2.1.1 Negative Refractive Index

It can be seen from Helmholtz's equation (2.14) that propagating waves exist in materials whether ϵ and μ are both positive or negative. In this equation, ϵ and μ enter as a product, so it would not appear to matter whether both the signs are positive or negative. Conventionally, we always express the refractive index n as $+\sqrt{\mu_r\epsilon_r}$ for positive materials ($\epsilon > 0, \mu > 0$).

But, one needs to be careful in taking the square root for $n = \pm\sqrt{\mu_r\epsilon_r}$, that is resolved by proper analysis. In real materials, the constitutive parameters (ϵ, μ) are complex quantities and causality requires the imaginary part to be positive since the materials are passive. For a left-handed material (LHM), the constitutive parameters satisfy

$$\epsilon_r = e^{i\phi_\epsilon}, \quad \phi_\epsilon \in (\pi/2, \pi]$$
$$\mu_r = e^{i\phi_\mu}, \quad \phi_\mu \in (\pi/2, \pi] \tag{2.20}$$

and now the refractive index can be expressed as

$$n = \sqrt{\epsilon_r\mu_r} = e^{i\phi_\epsilon/2}e^{i\phi_\mu/2} \tag{2.21}$$

and causality requires the imaginary part of the each $\sqrt{\epsilon}$ and $\sqrt{\mu}$ be positive

$$n = (\epsilon_r' + i\epsilon_r'')(\mu_r' + i\mu_r'') \tag{2.22}$$

where $\epsilon_r' = \cos(\phi_\epsilon/2)$, $\epsilon_r'' = \sin(\phi_\epsilon/2)$, $\mu_r' = \cos\left(\phi_\mu/2\right)$ and $\mu_r'' = \sin\left(\phi_\mu/2\right)$.

If the real part of each of the complex quantity in Eq. (2.22) goes to zero, the positive imaginary values result in a real $n < 0$. Therefore,

$$n = -\sqrt{\epsilon_r\mu_r}, \quad \epsilon_r < 0, \mu_r < 0 \tag{2.23}$$

A detailed derivation of it can be found in [11, 12].

2.1.2 Propagation of Waves in Left-Handed Material

For plane monochromatic wave proportional to $\exp(ikx - i\omega t)$, Eqs. (2.1)–(2.4) along with $\mathbf{D} = \epsilon\mathbf{E}, \mathbf{B} = \mu\mathbf{H}$ reduce to [13] (See Appendix A.2)

$$\mathbf{k} \times \mathbf{E} = \omega\mu\mathbf{H}$$
$$\mathbf{k} \times \mathbf{H} = -\omega\epsilon\mathbf{E} \tag{2.24}$$

It can be seen from these equations that for materials with $\epsilon > 0$ and $\mu > 0$ they form a right-handed triplet vectors and so the term *Right-Handed Materials* (RHM), and for materials with $\epsilon < 0$ and $\mu < 0$ they form a left-handed triplet vectors and so the term *Left-Handed Materials* (LHM).

This should not be confused with the left-handed chiral optical materials which are completely different. This has prompted authors to call these *backward wave media* [14], *negative phase velocity media* [12], *double negative media* [11] or *negative refractive index materials* [15].

The energy flux carried by the wave is determined by the Poynting vector **S**, given by

$$\mathbf{S} = \mathbf{E} \times \mathbf{H} \tag{2.25}$$

and the direction of the vector is given by the right-hand rule which is the same for right-handed materials. Since Poynting's Theorem is derived by considering conservation of energy [2], we can think of a matched interface of a right-handed material with a left-handed material and the energy flow has to be in the same direction. Therefore, according to Eq. (2.25) the vector **S** is in parallel with the wave vector **k** for right-handed materials and is anti-parallel for left-handed materials.

2.1.3 Propagation of Waves in Single Negative Medium

A single negative medium has either $\epsilon < 0$ or $\mu < 0$. For this analysis, we will assume $\epsilon < 0$ and $\mu = \mu_0$. The wave number k can be expressed as

$$k = \omega \sqrt{\mu_0 |\epsilon_r|} e^{i\phi_e/2} = k' + ik'' \tag{2.26}$$

By substituting k (2.26) in a plane wave ($\mathbf{E}(x, t) = \mathbf{E}_0 e^{i(kx-\omega t)}$), it can be expanded to

$$\mathbf{E}(x, t) = e^{-k''x} \mathbf{E}_0 e^{i(k'x-\omega t)} \tag{2.27}$$

The analysis for $\mu < 0, \epsilon > 0$ is similar. Therefore, the propagating wave in a single negative media is a decaying wave front.

2.2 Dispersion in Nonconductors

In the previous section, it has been shown that the propagation of EM waves in nonconducting media is governed by two properties of the material, which was assumed to be constant: the permittivity ϵ, and the permeability μ. It is well known from optics that the *refractive index* ($n = c\sqrt{\epsilon\mu}$) is a function of ω. Thus a prism bends blue light more sharply than red, spreading white light into a rainbow of colors. This phenomena is called *dispersion* and whenever the speed of a wave varies with frequency, the supportive medium is called *dispersive*. Although both ϵ and μ are function of ω, in practice, μ is very close to μ_0, for most natural materials, and its variation with ω is insignificant [16].

2.2.1 Lorentz Oscillator Model for Permitivity

The classical theory of the absorption and dispersion for nonconductors (insulators) is due mainly to Lorentz. The Drude model is applicable to free-electron metals. Although these models were based on classical ad hoc formulation, the quantum mechanical analogs are strikingly similar and to date, the Drude–Lorentz models are still very useful for developing a feel for optical properties of solids.

The Lorentz model considers an atom with electrons bound to the nucleus using a model described by a small mass tied to a large mass by spring. The motion of an electron is then described by

$$m\left[\ddot{\mathbf{x}} + \gamma\dot{\mathbf{x}} + \omega_0^2\mathbf{x}\right] = -e\mathbf{E}(\mathbf{x}, t) \tag{2.28}$$

where $m\gamma\dot{\mathbf{x}}$ is the damping force representing the energy loss mechanism which arises due to radiation from an atom due to different scattering mechanisms. The term $m\omega_0^2\mathbf{x}$ is the Hooke's law restoring force in the *electron oscillator* model. In the context of a classical model, there are two main assumptions in Eq. (2.28). The nucleus has been assumed to have infinite mass and the small magnetic force $-e\mathbf{v} \times \mathbf{B}/c$ on the electron due to the magnetic component has been neglected.

For a sinusoid electric field with frequency ω as $e^{-i\omega t}$, the displacement vector $\mathbf{x}$ is the solution of Eq. (2.28) and the dipole moment $\mathbf{p}$ contributed by each electron is [2]

$$\mathbf{p} = -e\mathbf{x} = \left[\left(\frac{e^2}{m_e}\right)\frac{1}{(\omega_0^2 - \omega^2) - i\omega\gamma}\right]\mathbf{E} = \xi(\omega)\mathbf{E} \tag{2.29}$$

where $\xi(\omega)$ is the frequency dependent atomic polarizability. Assuming a linear relationship between $\mathbf{p}$ and $\mathbf{E}$ due to small displacements, $\xi(\omega)$ is a complex quantity because of the damping term in the oscillator model.

For N atoms per unit volume, the macroscopic polarization is [4]

$$\mathbf{P} = N\langle\mathbf{p}\rangle = N\xi(\omega)\langle\mathbf{E}\rangle = \epsilon_0\chi(\omega)\mathbf{E} \tag{2.30}$$

where $\chi(\omega) = N\xi(\omega)/\epsilon_0$ is the frequency dependent complex susceptibility that is defined in relation to the constitutive parameters as

$$\mathbf{D} = \epsilon_0(1 + \chi(\omega))\mathbf{E} = \epsilon_0\epsilon_r(\omega)\mathbf{E} \tag{2.31}$$

Using Eqs. (2.29)–(2.31), the relative permitivity $\epsilon_r(\omega)$ is

$$\epsilon_r(\omega) = 1 + \frac{\omega_p^2}{(\omega_0^2 - \omega^2) - i\omega\gamma}, \qquad \omega_p^2 = \frac{Ne^2}{m_e\epsilon_0} \tag{2.32}$$

where ω_p is the plasma frequency of the insulator which is the oscillating frequency in Drude model where the restoring force term $m_e\omega\gamma$ is zero. In other words, the Drude model for metals is obtained directly from the Lorentz model for insulators simply by equating the restoring force to zero.

2.2.2 Anomalous Dispersion and Resonant Absorption

From the complex permitivity expression in Eq. (2.32), the real and imaginary part of $\epsilon_r = \epsilon_r' + i\epsilon_r''$ can be written as

$$\begin{aligned}
\epsilon_r'(\omega) &= 1 + \omega_p^2 \frac{\omega_0^2 - \omega^2}{(\omega_0^2 - \omega^2)^2 + \omega^2\gamma^2} \\
\epsilon_r''(\omega) &= \omega_p^2 \frac{\omega\gamma}{(\omega_0^2 - \omega^2)^2 + \omega^2\gamma^2}
\end{aligned} \tag{2.33}$$

where $\epsilon_r' = \mathtt{Re}[\epsilon_r]$ and $\epsilon_r'' = \mathtt{Im}[\epsilon_r]$. The imaginary part is directly associated with the absorption of the incident wave and also when the real part is negative. The frequency dependence of ϵ_r' and ϵ_r'' is plotted in Fig. 2.1. Except for a narrow region around the resonance, ϵ_r' increases with frequency called the normal dispersion. In the narrow region of the resonance it decreases with frequency called *anomalous dispersion*. This region is also the frequency interval of maximum absorption as will be shown later in this section. The width of this region is equal to the loss factor γ.

With ϵ now a complex function of ω, the dispersive medium admits x-polarized plane-wave solutions, as before,

$$\mathbf{E}(x,t) = \mathbf{E}_0 e^{i(kx-\omega t)} \tag{2.34}$$

However, the wave number $k = \omega\sqrt{\epsilon\mu}$ is complex, because ϵ is complex. Writing k in terms of real and imaginary parts of the refractive index

$$k = \beta + i\frac{\alpha}{2} \tag{2.35}$$

equation (2.34) becomes

$$\mathbf{E}(x,t) = \mathbf{E}_0 e^{-\alpha x/2} e^{i(\beta x - \omega t)} \tag{2.36}$$

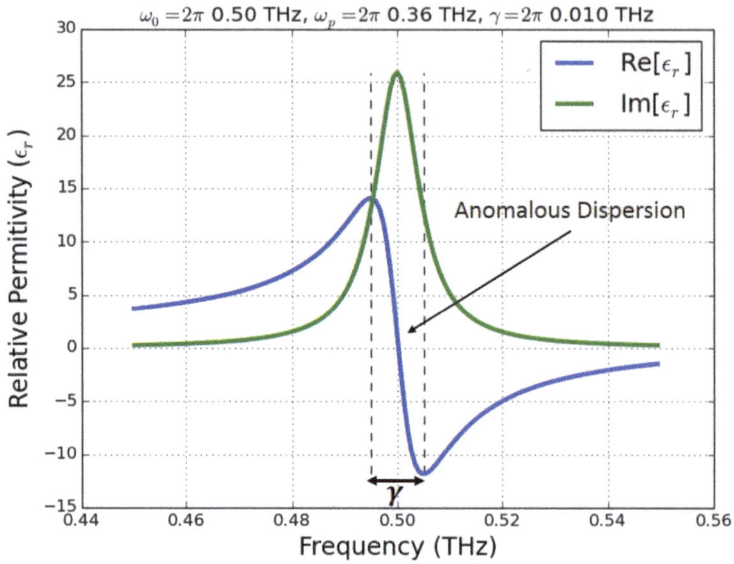

Fig. 2.1 Real and imaginary parts of the dielectric constant $\epsilon(\omega)/\epsilon_0$ in the neighborhood of a resonance. The region of anomalous dispersion is also the frequency interval for the maximum absorption

Evidently, $\alpha/2$ measures the *attenuation* of the wave. Because the *intensity* is proportional to E^2, it falls off as $e^{-\alpha x}$, α is called the *absorption coefficient*. For non-magnetic material ($\mu = \mu_0$), the complex frequency dependent refractive index n can be expressed as

$$n = n_R + i n_I = \sqrt{\epsilon_r' + i\epsilon_r''} \tag{2.37}$$

where the n_R and n_I are the real and imaginary part of complex n known as the *ordinary refractive index* and *extinction coefficient*, respectively.

Squaring and equating both sides of Eq. (2.37) we get

$$\epsilon_r' = n_R^2 - n_I^2, \qquad \epsilon_r'' = 2n_R n_I \tag{2.38}$$

From Eq. (2.38), the real and imaginary part of the refractive index are

$$\begin{aligned} n_R &= \left\{\frac{1}{2}\left[\sqrt{(\epsilon_r')^2 + (\epsilon_r'')^2} + \epsilon_r'\right]\right\}^{1/2} \\ n_I &= \left\{\frac{1}{2}\left[\sqrt{(\epsilon_r')^2 + (\epsilon_r'')^2} - \epsilon_r'\right]\right\}^{1/2} \end{aligned} \tag{2.39}$$

Now, the absorption coefficient α can be expressed in terms of the refractive index using the relation

$$k = n\omega\sqrt{\epsilon_0\mu_0} = k_0\omega(n_R + in_I), \qquad k_0 = \sqrt{\epsilon_0\mu_0} \tag{2.40}$$

From Eq. (2.40) we can express the absorption coefficient as

$$\alpha = 2k_0\omega n_I \tag{2.41}$$

Typically, for analyzing absorption around the resonance, n_I is sufficient indicator since ω does not change appreciably in that region.

Another optical parameter that provides independent information about the material in the frequency of interest is the reflection coefficient. From Eq. (2.39), the reflection coefficient at normal incidence is given by Wooten [4]

$$R = \frac{(n_R - 1)^2 + n_I^2}{(n_R + 1)^2 + n_I^2} \tag{2.42}$$

The plots for the real/imaginary part of the refractive index [Eq. (2.39)] and the reflection coefficient [Eq. (2.42)] are shown in Fig. 2.2. From the plots, we can see the implications of the frequency dependence of ϵ_r' and ϵ_r''. The plots show four distinct regions, transmission, absorption, reflection, and transmission again.

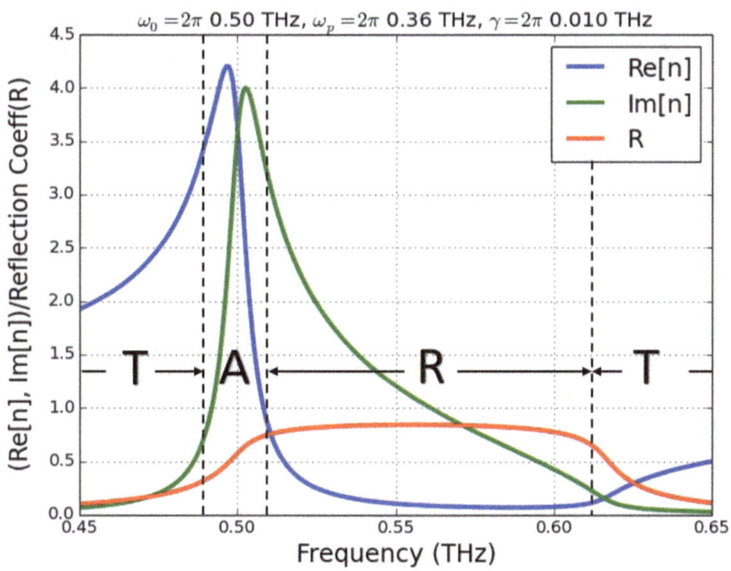

Fig. 2.2 Plots of the real and imaginary part of the refractive index [Eq. (2.39)] and the reflection coefficient [Eq. (2.42)] for an example material with $\omega_0 = 2\pi 0.5$ THz, $\omega_p = 2\pi 0.36$ THz and $\gamma = 2\pi 0.01$ THz

Before the onset of the resonance, the material behaves as a positive refractive index transmissive material. Then, for roughly the band around resonance ($\omega_0 \pm \gamma/2$), the material behaves as an absorber, reflective after that and for frequencies much higher than the plasma frequency, its transmissive again.

These plots give an insight on the considerations for building a metamaterial based modulator. By dynamically controlling the absorption peak, one can modulate an EM wave passing through the metamaterial around the resonance frequency.

2.3 Metamaterial as a Modulator

As explained in Sect. 1.2.2.2, the electrically coupled LC resonator (ELC) based metamaterial is suitable for planar design since the incident field can be incident to the normal of the device plane. For the ELC resonator shown in Fig. 2.3a, the average permittivity, ignoring the spatial dispersion, is of the Lorentz-like form [17–19]

$$\epsilon_{ELC} = \epsilon_a \left[1 - \frac{Ff^2}{f^2 - f_0^2 + i\gamma f} \right] \tag{2.43}$$

where ϵ_a is the permittivity of the background material, e.g. FR4, GaAs substrate, etc., $f_0 = 1/\sqrt{LC}$ is the resonant frequency in terms of its equivalent circuit parameters (Fig. 2.3b), γ is associated with the loss (R_{loss} in Fig. 2.3b), and F is associated with filing factor of the geometry of the unit cell. The real and imaginary part of the permittivity from Eq. (2.43) is plotted in Fig. 2.3c for an example design with $f_0 = 0.5$ THz, $\gamma = 0.01$ THz, and $F = 0.5$. The response is similar to the Lorentz oscillator model of non-conductors derived in Eq. (2.32). As observed in Fig. 2.2, the absorption of non-conducting materials is maximum in the vicinity

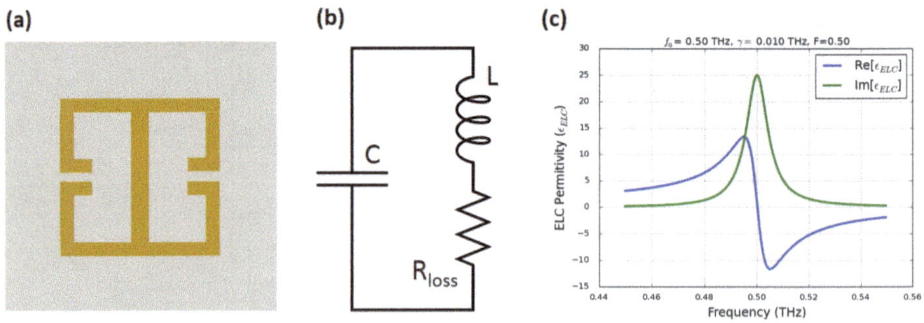

Fig. 2.3 (**a**) An ELC element used in most electric metamaterial design. (**b**) An equivalent circuit of the MM unit cell (**c**) Average permittivity (real and imaginary) of the MM [Eq. (2.43)] for an example design with $f_0 = 0.5$ THz, $\gamma = 0.01$ THz, and $F = 0.5$

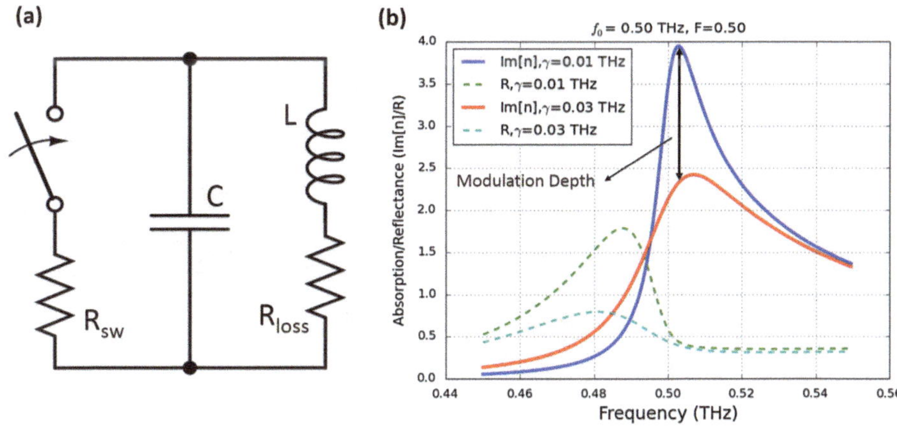

Fig. 2.4 (**a**) A conceptual circuit equivalent of a metamaterial based modulator (**b**) Absorption and reflectance plot of an ELC based metamaterial with and without the modulation resistance (R_{sw})

of molecular resonance frequency. It follows from that observation that, in order to design a metamaterial based modulator for a desired frequency, the resonance frequency of the metamaterial should be designed for that desired frequency and a dynamic method should be devised to either shift the resonance frequency or weaken the resonance by adding loss to the resonator.

One such method is shown conceptually in Fig. 2.4a where a resistive element is used to shunt the MM capacitance of the split gap to weaken the resonance. Assuming the shunt resistor changes the dissipation factor γ from 0.01 to 0.05 THz, the imaginary part of the complex refractive index [n in Eq. (2.39)], which corresponds to the absorption coefficient, is plotted in Fig. 2.4b. It can be seen from the plot that the absorption around the resonance frequency ($f_0 = 0.5$ THz) drops by approximately 40%. The reflectance plot shows little change around the modulation frequency. Terahertz modulators have been demonstrated based on this principle of shunting the split gap capacitance by optically pumping the substrate [20, 21] or electrically injecting carriers [22], a schematic of the basic structure shown in Fig. 1.10. This is the fundamental principle behind the metamaterial based terahertz modulator design in this work that is covered in depth in Chap. 4.

Metamaterial based terahertz modulators have also been demonstrated by dynamically controlling the capacitance or the inductance of the split gap capacitance. A dynamic terahertz metamaterial was realized by dynamically controlling the capacitance of the split gap by optically pumping the substrate and experimentally demonstrated by Chen et al. [23], a schematic of the structure shown in Fig. 1.9. The same work [23] also computationally demonstrated a dynamic terahertz metamaterial by controlling the inductance of the metamaterial unit cell as shown in Fig. 2.5.

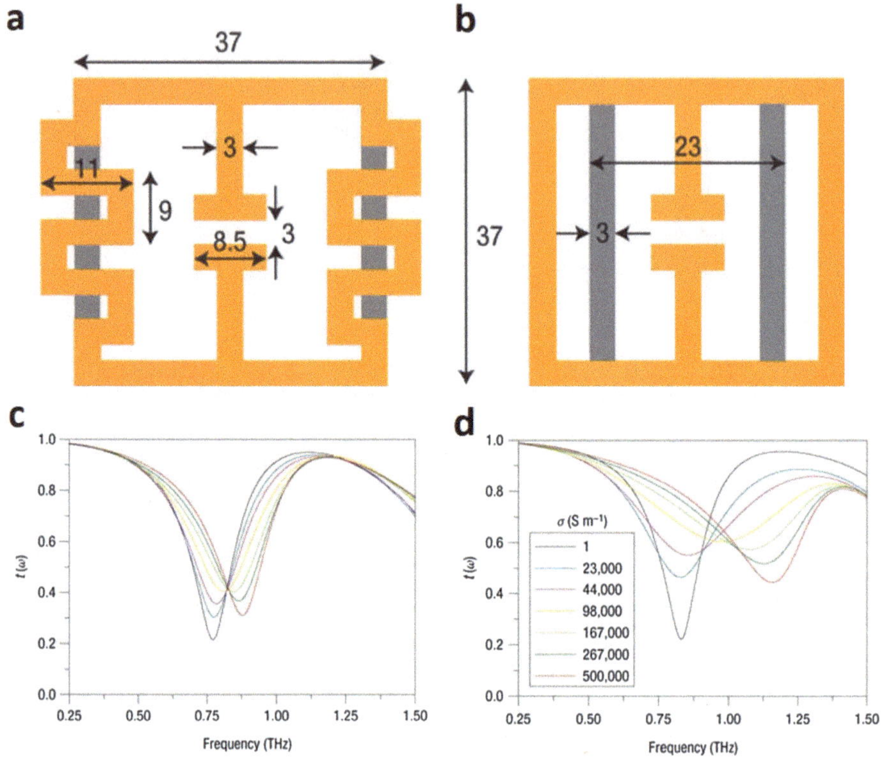

Fig. 2.5 An inductance tune dynamic terahertz metamaterial [23]. (**a**), (**b**) The photoexcited silicon regions form parallel current paths through the meandering loop sections (**a**) and across the regular loop sections (**b**) to effectively modify the inductance of the SRRs. The metal and silicon regions are displayed in *orange* and *grey*, respectively. The dimensions are shown in micrometers. (**c**), (**d**) Simulations of the structure of (**a**) and the structure of (**b**), both using the silicon conductivity values shown in the key

References

1. L.D. Landau, J.S. Bell, M.J. Kearsley, L.P. Pitaevskii, E.M. Lifshitz, J.B. Sykes, *Electrodynamics of Continuous Media* (Elsevier, Amsterdam, 1984)
2. J.D. Jackson, *Classical Electrodynamics*, 3rd edn. (Wiley, New York, 1998)
3. J.A. Kong, *Electromagnetic Wave Theory* (Wiley, New York, 1990)
4. F. Wooten, *Optical Properties of Solids* (Academic, New York, 2013)
5. S.A. Ramakrishna, T.M. Grzegorczyk, *Physics and Applications of Negative Refractive Index Materials* (CRC Press, Boca Raton, 2008)
6. S. Tretyakov, *Analytical Modeling in Applied Electromagnetics* (Artech House, Norwood, 2003)
7. G.V. Eleftheriades, K.G. Balmain, *Negative-Refraction Metamaterials: Fundamental Principles and Applications* (Wiley, New York, 2005)
8. C. Caloz, T. Itoh, *Electromagnetic Metamaterials: Transmission Line Theory and Microwave Applications: The Engineering Approach* (Wiley, New York, 2006)

9. N. Engheta, R.W. Ziolkowski, *Metamaterials: Physics and Engineering Explorations* (Wiley, New York, 2006)
10. J. Pendry, *Fundamentals and Applications of Negative Refraction in Metama* (Princeton University Press, Princeton, 2008)
11. R.W. Ziolkowski, E. Heyman, Phys. Rev. E **64**(5), 056625 (2001)
12. M.W. McCall, A. Lakhtakia, W.S. Weiglhofer, Eur. J. Phys. **23**(3), 353 (2002)
13. V. Veselago, Soviet Phys. Uspekh **10**(4), 509 (1968)
14. I.V. Lindell, S.A. Tretyakov, K.I. Nikoskinen, S. Ilvonen, Microwave Opt. Tech. Lett. **31**(2), 129 (2001)
15. W.J. Padilla, D.N. Basov, D.R. Smith, Mat. Today **9**(7–8), 28 (2006)
16. D.J. Griffiths, *Introduction to Electrodynamics* (Pearson, Boston, 2013)
17. R. Liu, T.J. Cui, D. Huang, B. Zhao, D.R. Smith, Phys. Rev. E **76**(2), 026606 (2007)
18. D.R. Smith, Phys. Rev. E **81**(3), 036605 (2010)
19. B.J. Arritt, D.R. Smith, T. Khraishi, J. Appl. Phys. **109**(7), 073512 (2011)
20. W.J. Padilla, A.J. Taylor, C. Highstrete, M. Lee, R.D. Averitt, Phys. Rev. Lett. **96**(10), 107401 (2006)
21. H.T. Chen, W.J. Padilla, J.M.O. Zide, S.R. Bank, A.C. Gossard, A.J. Taylor, R.D. Averitt, Optics Lett. **32**(12), 1620 (2007)
22. H.T. Chen, W.J. Padilla, J.M.O. Zide, A.C. Gossard, A.J. Taylor, R.D. Averitt, Nature **444**(7119), 597 (2006)
23. H.T. Chen, J.F. O'Hara, A.K. Azad, A.J. Taylor, R.D. Averitt, D.B. Shrekenhamer, W.J. Padilla, Nat. Photon **2**(5), 295 (2008)

Chapter 3
Experimental Methods

This chapter covers modeling, simulations and test techniques to successfully design and implement metamaterials for terahertz frequencies. Modeling and simulation of metamaterials is briefly covered in Sect. 3.1 with emphasis on the choice of boundary conditions to leverage the symmetry of metamaterial structures and thus dramatically reduce simulation time and solid-state memory requirements. Section 3.2 covers fabrication techniques using commercial foundry process with 45nm CMOS technology as an example. Few case studies on terahertz metamaterials are covered to showcase the power of using commercial semiconductor process for designing terahertz metamaterials. Finally, Sect. 3.3 covers test and characterization methods in detail to give the reader a solid background on characterizing terahertz metamaterials.

3.1 Electromagnetic Modeling and Simulations of Metamaterials

For most metamaterial design, predicting effective electromagnetic parameters accurately using analytic methods is not very practical. Therefore, computational methods for modeling and simulating metamaterials is an important and necessary step to design and implement metamaterials. Along with accurate geometrical modeling, the computational methods must be able to model several kinds of materials, including metals with finite conductivity and (possibly dispersive) dielectrics.

S. Rout, S. Sonkusale, *Active Metamaterials*, DOI 10.1007/978-3-319-52219-7_3

3.1.1 Boundary and Symmetry Conditions

A typical planar metamaterial contains a two-dimensional array of metamaterial unit cells. Direct simulation of the entire array is not computationally efficient. A more efficient approach, with a slight penalty in accuracy, is finding the solution of the Maxwell's equation for one unit cell with appropriate boundary conditions. After finding the solution for the unit cell, the complex scattering parameters S_{11} (reflection) and S_{21} (transmission) in the direction of propagation are computed. From the scattering parameters, refractive index (n), impedance (Z), effective permitivity (ϵ_{eff}), and effective permeability (μ_{eff}) can be computed [1–5].

If a mirror symmetry exists in the metamaterial structure that is being simulated, two oppositely located electric and/or magnetic boundary conditions maybe defined to emulate the entire array thus saving computation time and memory. The reason for this is that the boundary conditions work like mirrors and, when two mirrors are placed in front of each other, infinitely many mirror images of the unit structure are created. The designer has to ensure that in addition to geometric symmetry, field symmetry is also matched by the boundary condition.

The boundary condition of a metamaterial unit cell is shown in Fig. 3.1, where the unit cell is part of a metamaterial array in the X-Y plane. The excitation of the cell is provided by a linearly y-polarized electromagnetic wave propagating

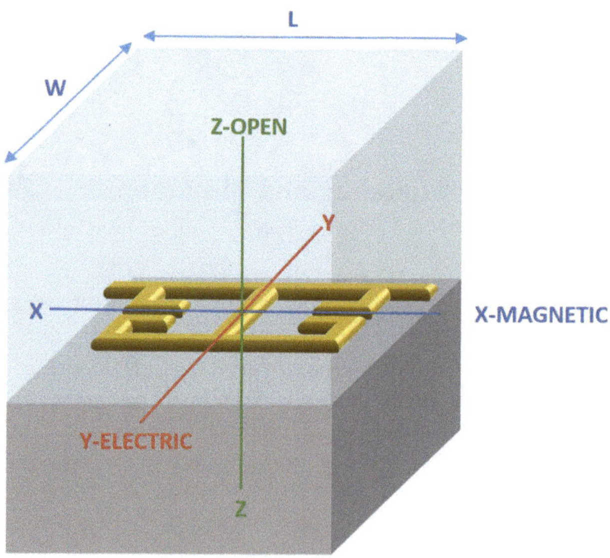

Fig. 3.1 Typical simulation method for metamaterials using unit cells. Unit cell showing electric/magnetic boundary conditions used in the Microwave Studio [6] simulations. Periodic boundary conditions are applied along the x-axis and y-axis to emulate the array of metamaterials in the X-Y plane. The direction of propagation of the electromagnetic field is along the z-axis, the electric field is oriented along the y-axis, the magnetic field along the x-axis

in the z direction. In order to mirror the unit cell, the two walls perpendicular to the y axis are defined as electric boundaries and the walls perpendicular to x axis are defined as magnetic boundaries. The top and bottom walls are defined as open boundaries for the EM wave propagation. Most EM solvers, like Microwave Studio [6], also allow "periodic" and "unit cell" boundary conditions that can be used to repeat unit cells in one, two, or three dimensions.

3.1.2 Homogenous Parameter Extraction

Extracting effective parameters ($\epsilon_{\mathrm{eff}}, \mu_{\mathrm{eff}}$) for metamaterials is a powerful tool to characterize the frequency dependence constitutive parameters.

For a homogeneous and isotropic material, the refractive index (n) and the wave impedance (z) can be expressed in terms of its scattering parameters (S_{11}, S_{21}) as [1–5]

$$n = \frac{1}{kd}\cos^{-1}\left[\frac{1}{2S_{21}}\right] \tag{3.1}$$

$$z = \sqrt{\frac{(1+S_{11})^2 - S_{21}^2}{(1-S_{11})^2 - S_{21}^2}} \tag{3.2}$$

where the wavenumber of the incident wave is $k = 2\pi/\lambda$ and d is the unit cell dimension. And the relationship between n and z to ϵ_{eff} and μ_{eff} is

$$n = \sqrt{\epsilon_{\mathrm{eff}}\mu_{\mathrm{eff}}}, \qquad z = \sqrt{\frac{\epsilon_{\mathrm{eff}}}{\mu_{\mathrm{eff}}}} \tag{3.3}$$

It is difficult from experiments to measure the scattering parameters but it is easily calculated from simulations and the full material response can be calculated using Eqs. (3.1)–(3.3).

3.2 Design for Fabrication in Foundry Processes

Since Pendry's demonstration of microwave artificial plasma using thin wire medium in 1996 [7], there has been an explosion of research activities in the area of metamaterials to artificially tailor materials properties for frequencies all the way up to visible spectrum [8, 9]. Of particular interest is the terahertz regime ("THz gap") that has numerous critical applications but lacks of natural materials responsive to that regime [10, 11] (see Sect. 1.1.1). As explained in Sect. 1.2.3, the scales of unit cells and the materials required to construct metamaterial in the terahertz regime make them the ideal candidate for conventional planar microfabrication.

Commonly used fabrication technique in the metamaterial community includes electron-beam lithography (EBL) [8] and nano-imprint lithography (NIL) [12]. EBL features very high resolution and can create patterns with feature size on the order of a few tens of nanometer [13]. However, EBL is very expensive and time consuming when used to fabricate large area metamaterial devices, which prohibit its usage for mass production. NIL is another option to fabricate nanometer feature size metamaterial. It is simple and features low-cost, high throughput [13]. But its throughput is very slow at the smallest resolution because its high resolution template patterning is performed by electron beam lithography. Additionally, it suffers from the problems of overlay, defects, and template wear [14]. All these factors render NIL unsuitable for mass production of metamaterial devices.

Commercial Gailum-Arsenide (GaAs) and Complementary Metal-Oxide Semiconductor (CMOS) foundry process are matured, low-cost, and suitable for very high volume production. Aggressive scaling of CMOS process to nanometer dimensions [15] assures its ability to meet the needs of metamaterial designs for short wavelengths down to far-infrared frequencies. Therefore, adopting designs to the vast variety of commercial foundry process will enable metamaterials to transition from research phase to commercial products that are mass producible.

3.2.1 *Typical 45nm CMOS Process*

It is beneficial to introduce the structure and dimensions of the commercial 45nm CMOS process. The commercial 45nm CMOS process usually has 5–12 metal layers. The 45nm CMOS technology employed in this work has 10 metal layers. The thickness of each metal layer and the dielectric between neighboring layers are fixed, which only limits the design freedom in vertical dimension when metamaterials are implemented on this CMOS process. The cross-section view of the metal and dielectric layers, along with their vertical dimensions, for the 45nm CMOS process is shown in Fig. 3.2.

There are 10 metal layers in this CMOS process, which are termed Metal-1 to Metal-10 in Fig. 3.2a. The thickness of each metal and dielectric layer is named after the letter in the parenthesis and defined in table shown in Fig. 3.2b. Range of values shown captures the entire feasible range of values across multiple foundries.

In general the lower level metal layer and dielectric spacer are thinner while the upper lever metal layer and dielectric spacer are thicker. The thickness of metal layer and dielectric spacer may vary 10–20% due to foundry process variation.

The bottom layer in CMOS is usually a silicon handle wafer layer. It can be etched by post processing. The next layer on top of the silicon handle wafer layer is the bulk silicon layer which carries the transistors. In this work, the bulk silicon is assumed to be etched down to 200$\sim$1000 nm.

There are design rules required by the CMOS foundry limiting the minimum width of metal traces and spacing between metal traces on each metal layer. The lower metal layers have smaller minimum feature sizes while the upper metal layers

(a)

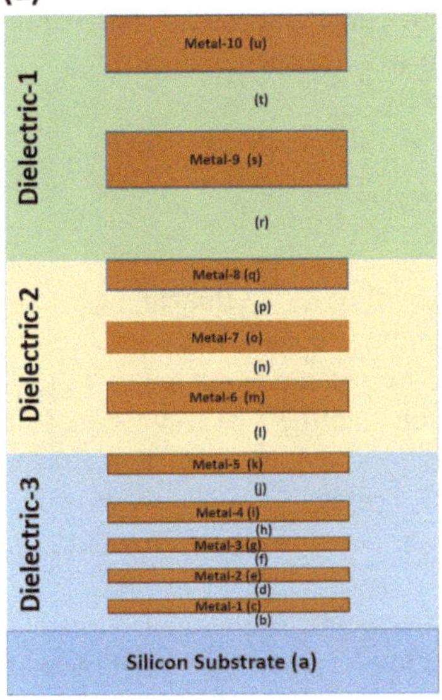

(b)

Name of layer	Value of thickness(nm)
a	200~1000
b	100~200
c	100~200
d	50~200
e	100~200
f	50~200
g	100~200
h	50~200
i	150~250
j	150~250
k	150~250
l	250~350
m	250~350
n	250~350
o	250~350
p	250~350
q	250~350
r	600~1000
s	1000~1400
t	1300~1900
u	1000~1400
x	2500~7000
y	800~2200
z	700~2000

Fig. 3.2 (**a**) The *cross-section view* of CMOS metal stack. The thickness of each metal and dielectric layer is named after the letter in the parenthesis and defined in table in (**b**)

have larger minimum feature sizes. So the design strategy employed in this work is to implement the higher frequency metamaterials on the lower metal layers and implement lower frequency MMs at higher level metal layers.

Another important design rule is the metal fill density rule on each metal layer in CMOS. Typically, the CMOS foundries require us to keep the metal fill density around 50% on each metal layer for high yield of fabrication. This will invariably restrict the kind of metamaterial geometries that can be implemented. For example, babinet metamaterial which has continuous metal layers will violate antenna rules and is not feasible.

3.2.2 *Physical Properties of Metal and Dielectrics at Optical Frequencies*

Conductive material, such as metals have different responses to the low frequency RF/microwave radiation and IR/optics radiation. At low frequencies, metals attenuate all the incident RF/microwave radiation that is impinged on its surface because of its high conductivity. At IR/optical frequencies, however, metals are no longer perfect conductors and the conductivity of metal is frequency dependent. Therefore, the response of metal to incident IR/optical radiation is also frequency dependent.

Metal is not a source free medium and is filled with free charge and currents. The relation between the electric flux density **D**, current density **J**, and the induced curl of magnetic field **H** when an EM radiation with frequency ω is shined onto the metal can be expressed as [16]:

$$\nabla \times \mathbf{H} = j\omega \left(\epsilon + \frac{\sigma}{j\omega} \right) \tag{3.4}$$

The term $\epsilon + \frac{\sigma}{j\omega}$ is defined as effective permittivity ϵ_e that includes both the dielectric and conductive properties [16]. It is more convenient for us to describe the response of metal to IR/optics using ϵ_e. The second term in ϵ_e, which is $\frac{\sigma}{j\omega}$, is frequency dependent [16]. It is much greater than the ϵ at low frequencies and that is why metals behave like perfect conductors at low frequencies. When the frequency of incident radiation approaches IR/optics, $\frac{\sigma}{j\omega}$ starts to drop and at a certain frequency, it becomes zero and the metal behaves like a dielectric medium with permittivity ϵ.

The most suitable model to describe the frequency dependent permittivity of metal at IR is Drude model [16]. The Drude model uses classical kinetic theory to describe the macroscopic behavior of electrons in solids where the EM radiation is impinged on [17].

Using Drude model the *dispersivity* (frequency dependence) of the permitivity can be expressed as [18]:

$$\epsilon_e = \epsilon_1 + i\epsilon_2 = \epsilon_\infty - \frac{\omega_p^2}{\omega^2 + i\omega\omega_\tau} \tag{3.5}$$

where ϵ_1 and ϵ_2 are the real and imaginary part of ϵ_e, respectively. ω_p is the plasma frequency, ω_τ is the damping frequency.

The Drude model, along with experimental data, of Al and Cu, which are the most commonly used metal in CMOS and, Gold which is mostly used in GaAs process, is shown in Fig. 3.3 [18].

The most common dielectrics used in CMOS processes include benzocyclobutene (BCB), silicon dioxide (SiO_2), tetraethoxysilane (TEOS), with polymide as the passivation layer. Silicon nitride (SiN_3) is mostly used in GaAs process. These dielectrics are quite suitable for terahertz devices but one will have to accurately characterize these materials for their frequency disperson which then needs to be incorporated in the design flow

3.2.3 *Case Studies*

In this section, we will explore numerical simulation based case studies utilizing CMOS process using the metal stack shown in Fig. 3.2 to design metamaterials for various applications.

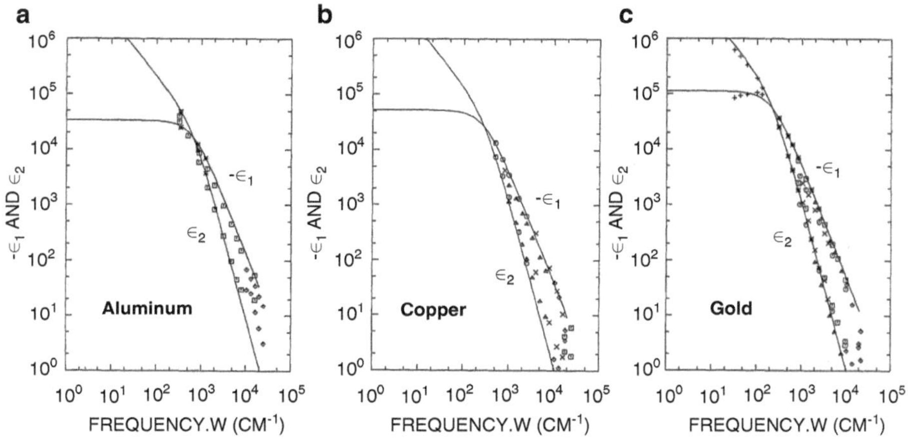

Fig. 3.3 Experimental data vs. Drude Model (*solid lines*) of the real and imaginary part of the permitivity for (**a**) Aluminum (**b**) Copper and (**c**) Gold, most used metals in CMOS and GaAs foundry process [18]

3.2.3.1 Single Layer Metamaterial Operating at 100 μm Wavelength

The first case investigated is a single layer metamaterial consisting of ELC resonators implemented on the top layer metal M10. ELC resonators exhibit strong coupling to applied electric field and negligible coupling to the magnetic field [19]. ELC resonators have been used as a negative index material [20], dual-resonant terahertz metamaterial [21], and high-speed terahertz modulator [22].

Its dimensions have been carefully tuned to achieve the desired EM response to incident wave. Its dimensions are depicted in Fig. 3.4a. The electric field inside and around the ELC resonator at the operating frequency is shown in Fig. 3.4b, emphasizing the concentrated electric field in the split gap at resonance. The surface current at the operating frequency is shown in Fig. 3.4c. The transmissivity and reflectivity of the metamaterial is shown in Fig. 3.4d. At its operating frequency of 2.835 THz, transmissivity T reaches −25 dB and the reflectivity R is almost 0 dB.

This ELC resonator based metamaterial can be used as a spatial band-stop at THz frequencies. A negative index metamaterial can be achieved if it is stacked on top of magnetic metamaterials such as SRR resonator [20]. The operating wavelength of this ELC resonator based metamaterial can be decreased by scaling down its dimensions and the lower limit of its operating wavelength is constrained by the minimum line spacing and width in this metal layer. Additional protection layers on top of the metal layers, typically polymide, fabricated by foundries may be lossy and dampen the metamaterial resonance.

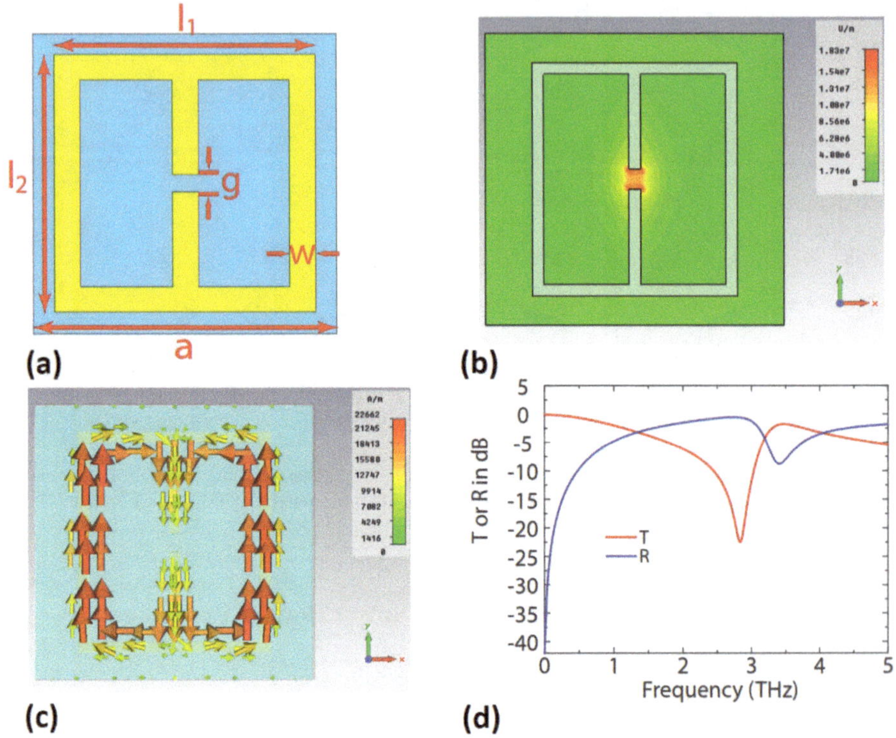

Fig. 3.4 EM simulation results of ELC resonator in 45nm CMOS process. (**a**) Dimensions of an ELC resonator metamaterial implemented on the top metal layer M10 in CMOS metal stack. All units are in μm. $l_1 = 14$, $l_2 = 16$, $g = 1.4$, $w = 0.8$, $a = 20$ (**b**) Electric field distribution of the resonator. (**c**) Surface current distribution (**d**) Simulated transmissivity T and reflectivity R of the metamaterial

3.2.3.2 Multi-Layer Metamaterial Design

One of the advantages of using CMOS process is access to multiple layers of metamaterial, so that you can implement multi-layered metamaterial designs. In this case, an ELC resonator and SRR resonator are implemented on the Metal-10 (M10) and Metal-9 (M9) layers, respectively. They have distinct resonant frequency and form a multi-band metamaterial collectively. The three-dimensional view of this two-layer metamaterial design and its dimension are shown in Fig. 3.5.

The transmissivity T and reflectivity R of SRR on M9 alone is shown in Fig. 3.6a. The resonant frequency of SRR on M9 is located at 1 THz. The cut-wire effect cause another dip in T at around 3.5 THz. The simulated transmissivity T and reflectivity R of the ELC resonator on M10 is shown in Fig. 3.4d. Its resonant frequency is located at 2.835 THz. The transmissivity T and reflectivity R of the stack of ELC on M10 and SRR on M9 are shown in Fig. 3.6b.

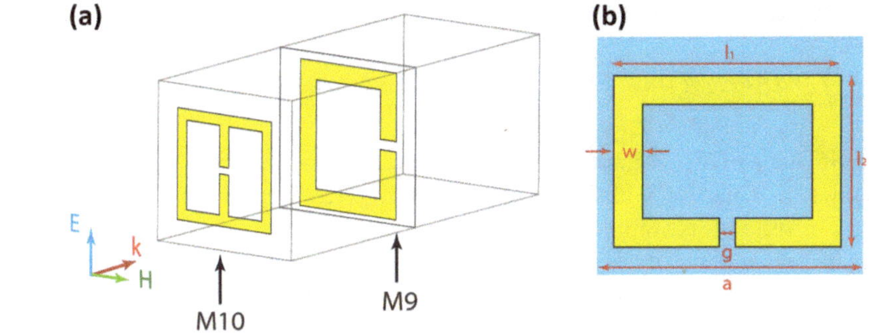

Fig. 3.5 (**a**) 3D view of a two-layer metamaterial composed of ELC resonator on M10 layer and SRR resonator on M9 layer (**b**) Dimensions of SRR resonator metamaterial implemented on the metal layer M9 in CMOS metal stack. All units are in μm. $l_1 = 19$, $l_2 = 18$, $g = 0.8$, $w = 0.8$, $a = 22$

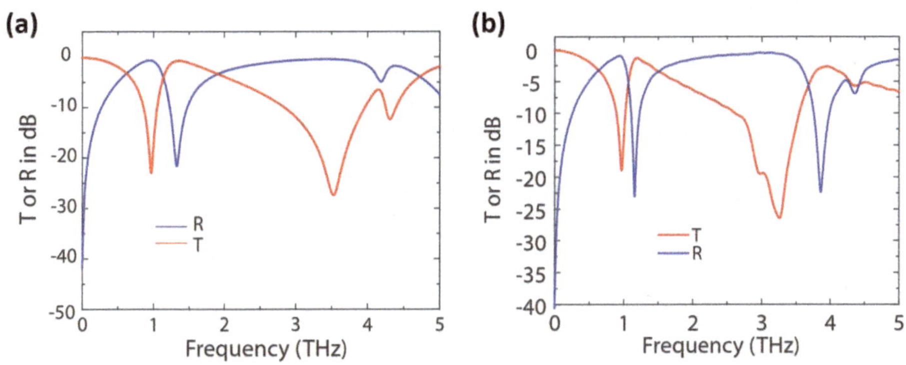

Fig. 3.6 Transmissivity T and reflectivity R of (**a**) SRR resonator only on M9 and, (**b**) of two-layer metamaterial composed of ELC resonator on M10 and SRR resonator on M9

As depicted in Fig. 3.6b, there are two operating frequency bands where T is very low and R is very high. These two frequency bands are around 1 THz and 3.2 THz, respectively. The resonance at 1 THz is contributed by SRR on M9 layer. The resonance at 3.2 THz is caused by the combination of resonance of ELC on M10 and the cut-wire effect of SRR on M9.

The alignment of two-layers of metamaterial is non-trivial. Misalignment may cause additional coupling that corrupts the desired dual-band response. CMOS foundry has the advanced technology to ensure the alignment of more than ten metal layers. It is very difficult to achieve in the university clean room environment. The dual-band frequency response of this metamaterial can be utilized to build a dual-band THz antenna featuring compact size and high throughput. It can also be used in hyper spectral imaging system.

3.3 Test and Characterization

3.3.1 Terahertz Time-Domain Spectroscopy (THz-TDS)

The significance of the terahertz, or far-infrared, region of electromagnetic spectrum was shown in Chap. 1, with applications ranging from astronomy to biological sciences. In spite of its significance, terahertz spectroscopy devices were in infancy till the late 1980s primarily due to the difficulty of building devices in that frequency regime. Electronic sources and detectors for millimeter and sub-millimeter waves were limited to about 100 GHz. Fourier Transform Infrared (FTIR) spectroscopy, on the other hand, was useful only above 5 THz due to lack of powerful incoherent sources. The advent of lasers motivated research in far-infrared generation using the difference-frequency generation in non-linear crystals [23], which served as the seed for the development of terahertz time-domain spectroscopy (THz-TDS) [24–28], which still serves as a system of choice for high dynamic range THz spectroscopy. The THz-TDS system generated a great deal of interest in THz imaging after first images acquired using the THz-TDS was reported in 1995 [29].

THz-TDS system is based on single-cycle bursts of electromagnetic transients generated opto-electronically using femtosecond duration laser pulses. Typically, the single-cycle burst are of 1 ps duration whose spectral density spans from near 100 GHz to above 5 THz. Optically gated detection allows direct sampling of the generated terahertz wave with sub-picosecond accuracy. From this measurement, the real and imaginary part of the dielectric function of the medium can be extracted without the use of Kramers-Kronig relation (Eq. (2.11)).

Compared to other THz spectroscopic systems, THz-TDS systems generally offer higher signal-to-noise ratio, a wide spectral band (0.1–5 THz), and faster acquisition time.

3.3.1.1 Terahertz Time-Domain Spectrometer

Figure 3.7 shows the schematic of a terahertz time-domain spectrometer. It consists of a femtosecond laser source, a beam splitter to split the laser beam into two path, one for the photoconductive THz transmitter and the other (through the optical delay line) for the optically gated photoconductive THz receiver. A set of collimating optics is used to focus the THz wave onto the sample and the THz receiver. An optical copper is used in the transmitting laser path which along with the lock-in amplifier serves as coherent mixer to eliminate DC and low-frequency noise in the system. The computer controls the delay line and measures the amplified, lock-in detected photocurrent versus path length. After capturing the entire THz pulse, the desired spectral response is generated using the Fast Fourier Transform (FFT). In the next few paragraphs we will describe the function of each component of the spectrometer.

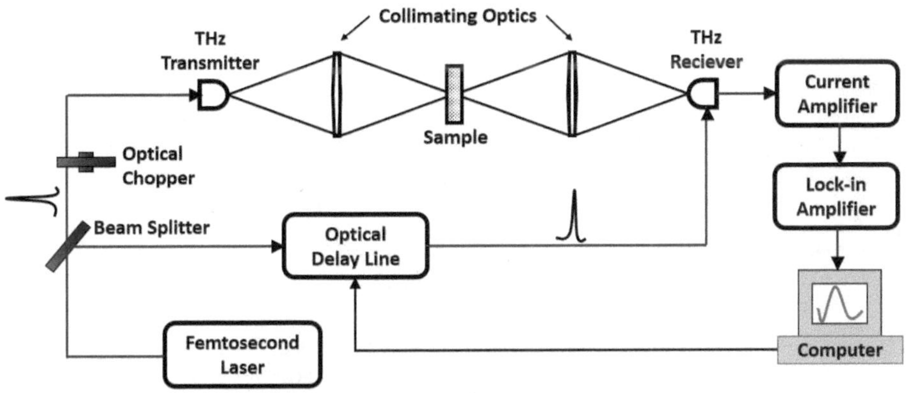

Fig. 3.7 Schematic of a typical THz time-domain spectrometer using a femtosecond laser source and photoconductive THz transmitter and receiver

3.3.1.2 Laser Sources

Solid-state laser sources like the Ti:Sapphire laser [30] with wavelength around 800 nm have largely displaced the older dye lasers [31] and they are a mature and commercial product now. The typical repetition rate of these lasers is about 100 MHz with 10–50 mW of laser power that is sufficient for most of the THz transmitters and receivers. Recent development of all-solid-state femtosecond lasers [32] will result in lasers that are more practical, compact, and cost-effective.

3.3.1.3 THz Transmitters and Detectors

Both the THz transmitter and receiver are based on the photoconductive ("Auston") switch [26, 33], which consists of a semiconductor bridging the gap in a transmission line or a dipole antenna structure as shown in Fig. 3.8a. The response of the voltage biased photoconductive antenna to a pulsed laser source is shown in Fig. 3.8b. The current through the switch rises rapidly at the incidence of the laser pulse and decays at a rate proportional to the carrier lifetime of the bridging semiconductor, generating a picosecond-rate transient current $J(t)$. The transmission line or the dipole antenna is designed such that this transient current is radiated into the free space.

To convert the photoconductive switch to a receiver, a trans-impedance amplifier (current-to-voltage converter) is connected across the switch instead of the voltage bias. The electric field in the bridging gap is now provided by the incident THz pulse. The current only flows through the receiving switch, the photodetector, when carriers are generated by the incident laser source resulting in photoconductive sampling of the THz pulse as shown in Fig. 3.8c. Since electronics are not fast enough to sample the picosecond THz pulse directly, repetitive measurement of the

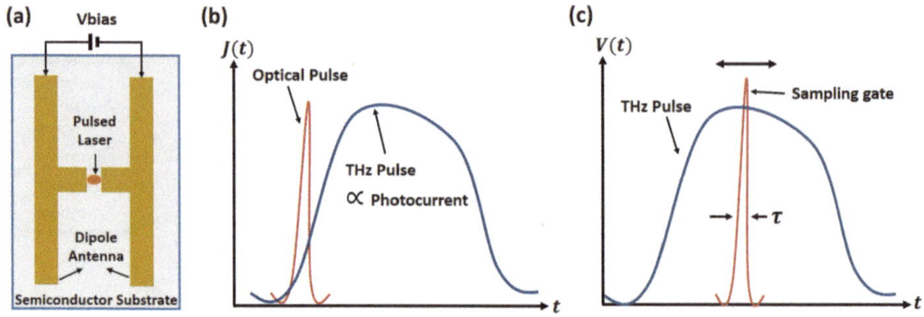

Fig. 3.8 (**a**) Photoconductive "Auston" switch integrated in a THz dipole on a substrate. Laser pulses are focused onto the gap in the dipole antenna which inject carriers into the substrate, which are swept across the dipole due to the bias voltage across. (**b**) Typical current response $J(t)$ of a photoconductive switch to a short laser pulse. (**c**) Photoconductive sampling measures the THz pulse $v(t)$ within the sampling interval τ

photoconductive sample is used for each sample point. If the photo-carrier lifetime, τ, is much shorter than the THz pulse, the photoconductive switch acts as a sampling gate for the THz pulse. Since the laser pulse which triggers the transmitter and gate the detector originate from the same source, the relative delay can be controlled through a optical delay line (Fig. 3.7) and the entire THz pulse is sampled step by step.

Although transmitter and receiver structures are similar in principle, photoconductive switches are designed specifically for transmitters [26, 34, 35] and receivers [26, 36] for better efficiency.

3.3.1.4 Bandwidth Limitation of THz Detectors

The photocurrent transient of the detector $J(t)$ can be expressed as the following convolution:

$$J(t) = \int \sigma(t - t')E(t')dt' \tag{3.6}$$

where $\sigma(t)$ is the transient photoconductivity and $E(t)$ is the incident electric field of the THz pulse. $E(t)$ is faithfully reproduced by $J(t)$ when the photoconductive transient is much shorter than the THz pulse. The high-frequency response of the detector is determined by the photoconductive transient and the low-frequency response is determined by the antenna structure.

The photocurrent decay time needs to be less than 0.5 ps in order to resolve transients in the THz regime. The intrinsic carrier lifetime in semiconductors is far slower, therefore the carrier lifetime needs to be reduced below its intrinsic value, commonly achieved by introducing defect states that have a fast carrier capture rate, either during crystal growth or afterwards through ion implantation.

Low temperature grown GaAs (LT-GaAs) [37] and radiation-damaged silicon-on-sapphire (RD-SOS) [38] are two such examples.

The electric field across the photoconductor can differ from the incident THz field due to the frequency response of the antenna structure.

3.3.1.5 Collimating and Focusing Optics

An important part of both the transmitter and receiver photoconductive antenna is the substrate lens. Because of the presence of the substrate on one side of photoconductive antenna, a substrate lens is required to efficiently couple the THz radiating pulse from the dipole to the free space and also magnifies the dipole antenna, increasing the antennas efficiency as well [39].

In order to incorporate the THz source and detector antennas into a spectroscopy system, a THz optical system required that can guide the THz wave from the source to the detector and focus it on to a diffraction-limited spot at the sample, and such a system is shown in Fig. 3.9. The THz radiation which is coupled into free space using substrate lenses is collimated and focused using a pair of off-axis parabolic mirrors (OAPM). This parallel, diffraction-limited collimated beam can be focused to a diffraction-limited spot at the sample using a pair of focusing lenses or OAPMs (dashed box in Fig. 3.9).

Although the OAPMs are difficult to align, they offer high reflectivity and achromatic operation over the entire THz range and visible lasers can be used for alignment. Fused quartz lenses can be used as an alternative for frequencies below 1 THz and up to 10 THz with silicon lenses but alignment with visible light or laser is not possible. Another useful lens is TPX (poly-4-methyl-pentene-1), a polymer which has low absorption and dispersion throughout the THz range [40], but is difficult to polish for their softness.

Alignment of OAPMs is covered in detail in Sect. 3.3.3.

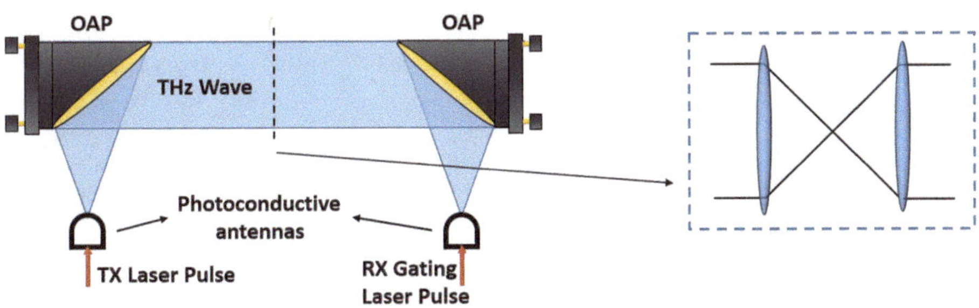

Fig. 3.9 The THz radiation is coupled in and out of the photoconductive antennas using substrate lenses and collimated using a pair off-axis parabolic mirrors (OAPM). Optionally, a pair of lenses or OAPMs can be used to focus the THz wave to a diffraction-limited spot at the sample (*dashed box*)

3.3.1.6 Lock-In Detection

Lock-in detection is primarily used when the desired signal is buried in the noise of the measurement system and cannot be measured directly. The received signal to be measured, $V_{\text{signal}}(t)$, is typically contaminated by offset and noise that can be expressed as [41]

$$V_{\text{signal}}(t) = V_{\text{IN}} + V_{\text{offset}} + V_{\text{white-noise}}(t) + V_{\text{flicker-noise}}(t) \tag{3.7}$$

where V_{IN} is the desired DC signal to be measured, e.g. optically gated THz electric field, V_{offset} is the DC offset voltage present both in the optics and electronics of the THz-TDS system, $V_{\text{white-noise}}$ is the *white gaussian* noise in the system which can be eliminated by simple averaging, and $V_{\text{flicker-noise}}(t)$ is the *flicker noise* which is inversely proportional to frequency ($1/f$). Even if we can remove the DC offset voltage by some calibration method and eliminate the white-noise by averaging, the measured signal will still be contaminated by the low-frequency drift associated with the flicker-noise. One powerful technique to remove the offset and the drift is by intentionally chopping the signal "on" ($V_{\text{IN-ON}}$) and "off" ($V_{\text{IN-OFF}}$) and taking the difference ($V_{\text{IN-ON}} - V_{\text{IN-OFF}}$). It can be seen from Eq. (3.7) that the offset voltage can be eliminated as well as low-frequency drift that is lower in frequency than the chopping frequency. And if we average the signal after the difference, the white-noise can also be eliminated, leaving us with only the noise contribution at the chopping frequency. Although this method can be implemented using a combination of analog and digital electronics, an elegant and powerful way is to use a lock-in amplifier.

A typical lock-in detection application is shown in Fig. 3.10 in which a laser is used to illuminate a sample and a detector, along with optics, is used to measure the emitted fluorescence. The laser is chopped at a particular frequency so the fluorescence turns "on" and "off" at that frequency. The same chopping frequency is provided as reference to the lock-in amplifier for it to "lock" and determine the operating frequency. Typically, the reference is converted to a sinusoid using a voltage-controlled oscillator (VCO) [42] with an adjustable phase which is multiplied with the input signal and time averaged to form the output voltage V_{out}

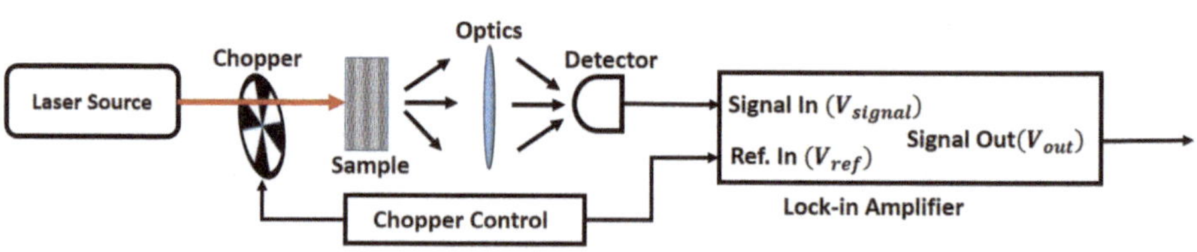

Fig. 3.10 A typical lock-in amplifier application in which one measures the fluorescence from a sample that is illuminated by a chopped laser source [41]

$$V_{out} = \langle V_{signal}(t) \cos(\omega t + \phi) \rangle \quad (3.8)$$

This method will pick out the fundamental Fourier component of the signal, $V_{signal}(t)$, which is proportional to the desired DC value, V_{IN}. This multiplying function is carried out by analog or digital electronics called mixer or a phase sensitive detector (PSD) [42]. Lock-in systems are sometimes referred to as special purpose *correlators* [43]. Most lock-in amplifiers are dual phase systems that incorporate a pair of PSDs which enables them to measure both the *in-phase* and *quadrature* component of the coherent input signal [43], allowing them to measure the magnitude and phase of the input signal.

3.3.1.7 Terahertz Time-Domain Data Analysis

As with most spectroscopic techniques, THz-TDS requires two measurements: one reference waveform $E_{ref}(t)$ measured without the sample or with sample with known dielectric properties and, the second waveform $E_{sample}(t)$, in which the radiation interacts with the sample. In order to observe the spectrum of the measurement, $E_{ref}(t)$ and $E_{sample}(t)$ can directly be Fourier-transformed to obtain the respective spectra, $E_{ref}(\omega)$ and $E_{sample}(\omega)$. One such example, where the absorption properties of methamphetamine (MA) is studied in the frequency range of 0.2–2.6 THz [44], is shown in Fig. 3.11. An useful trick which helps improve the readability of the measurement is a differential spectra that can be expressed as

$$\Delta E(\omega) = \frac{E_{sample}(\omega) - E_{ref}(\omega)}{E_{ref}(\omega)} \quad (3.9)$$

This technique is particularly useful when the difference between the sample and reference is very small.

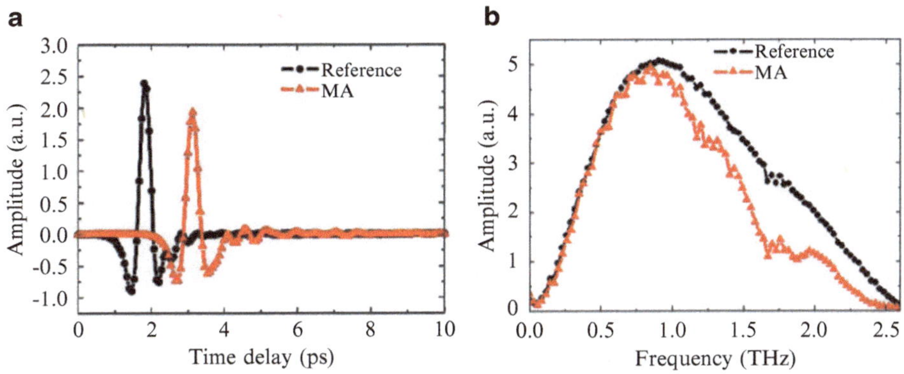

Fig. 3.11 (**a**) THz time-domain transmissive plot of the reference (no sample) and the sample, MA (methamphetamine) and (**b**) corresponding THz frequency-domain spectra [44]

Typically, the time-domain data is modified before performing the Fourier-transform on it, mostly to remove reflections that are prominent when a sample is thick. A thick sample is defined as one for which the transit time of the THz pulse is much larger than its duration. Semiconductor samples typically have their substrate thickness ranging from 100–500 μm, making them a thick medium. Because the measurement is done in time-domain, substrate reflections can be windowed out of the raw data without much loss in spectral resolution or accuracy. The Fresnel coefficients for transmission and reflection are used to remove the artifacts from multiple reflections at the boundary of the sample [39].

3.3.2 *Continuous-Wave (cw) Terahertz Spectroscopy*

Although THz-TDS have their advantages, continuous-wave, or frequency-domain, terahertz spectrometers are becoming popular because of their low-cost and ease of operation. Additionally, they offer the advantage of higher resolution with precise frequency control, compact footprint since a delay stage is not necessary and typically they use diode lasers which remain more cost-efficient than pulsed lasers used in THz-TDS.

3.3.2.1 A Continuous-Wave Terahertz (cw-THz) Spectrometer

Optical heterodyning is becoming a established standard for generating continuous-wave terahertz wave [45], and distributed feedback (DFB) lasers are becoming the standard in optical source because of their compact size, cost-effective, and hassle-free solution. A typical continuous-wave terahertz (cw-THz) spectroscopy setup is shown in Fig. 3.12 as implemented in one of the commercial products, Toptica

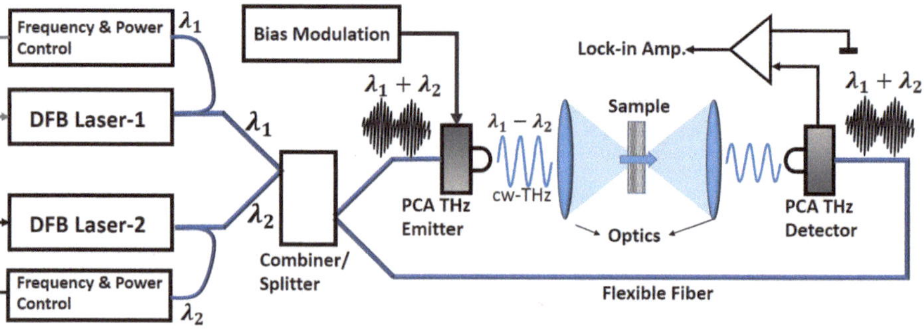

Fig. 3.12 Schematic of a typical continuous-wave terahertz (cw-THz) spectrometer. The continuous terahertz wave is created using the beat frequency of two distributed feedback (DFB) lasers that is transmitted and detected using photoconductive antennas

TeraScan [46]. Using interferometric frequency control [47], two DFB laser diodes are tuned to adjacent wavelengths λ_1 and λ_2 (e.g., 853 nm and 855 nm). The two-lasers are superimposed using a combiner and optionally, the two-color laser can be amplified for higher power. Using a 50:50 fiber splitter, two fiber outputs are created to irradiate the transmit and receive photomixers made from interleaved metal-semiconductor-metal structures. Resonant antenna around the photomixers help in radiating and receiving the terahertz wave of the beat frequency $(\lambda_1 - \lambda_2)$. The bias is usually modulated to function as the optical chopper in the lock-in detection as explained in Sect. 3.3.1.

3.3.2.2 Laser Sources

Distributed feedback (DFB) lasers have established themselves as a compact, cost-effective, and carefree alternative to traditional lasers like the Ti:Sapphire laser [30]. DFB lasers at 0.8 μm have become the preferred choice for GaAs photomixers [46–50], because of their high thermal tuning coefficient of ∼25 GHz/K per diode [51], resulting in about 2000 GHz scanning frequency. DFB diodes feature a grating structure within the active region of the semiconductor [48] (see Fig. 3.13a), restricting their emission spectrum to a single longitudinal mode. Varying the grating pitch either thermally or electrically tunes the lasing wavelength, and continuous frequency scans of 2000 GHz can be realized by means of a temperature sweep of 50 K. For frequency stabilization as well as controlled frequency tuning, each DFB laser is thermally and electrically regulated by an electronic feedback as shown in Fig. 3.12. The temperature is mostly controlled by a thermo-electric cooler (TEC) [52], and the frequency of the laser is measured using a fiber-coupled quadrature interferometer [47] (see Fig. 3.13b). Alternatively, instead of using a feedback scheme, the thermo-electric control is calibrated and values stored in a look-up table that can be used to control the DFB laser, based on frequency sensed from the quadrature interferometer [52].

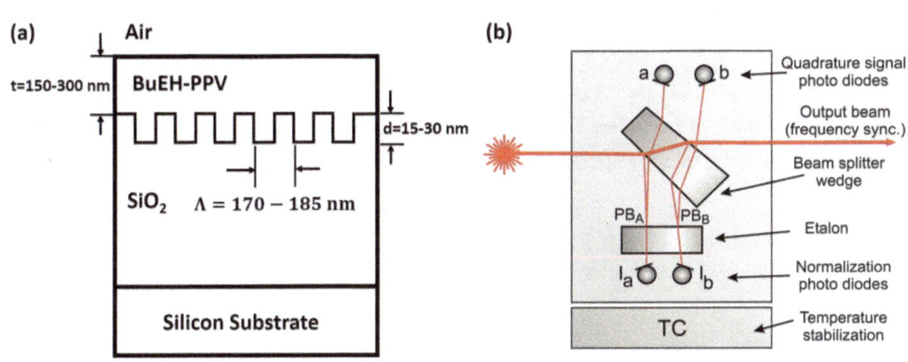

Fig. 3.13 (**a**) Schematic of a polymer DFB laser (not drawn to scale) [48]. (**b**) Quadrature interferometer for frequency control of DFB lasers [47]

3.3.2.3 THz Transmitters and Detectors

Typically, a transmitting photomixer consists of an epitaxial layer of low-temperature-grown gallium arsenide (LTG-GaAs) with interdigitated electrodes fabricated on a semi-insulating GaAs substrate. Two-lasers illuminating the gaps between the electrodes generate difference frequency photocurrents, and terahertz power is generated as the photocurrents drive the integrated antenna. The advantages of LTG-GaAs are its short photoconductive lifetime ($\tau < 1\,\text{ps}$) and its high electrical breakdown ($E_B > 5 \times 10^5\,\text{V} \cdot \text{cm}^{-1}$) [45]. New generation of terahertz sources and detectors are made up of high-energy ion-implanted GaAs interdigitated photomixers [46], which offer the advantage of highly reproducible photoconductive materials. The photomixers are integrated within self-complementary log-periodic spiral [46] or linear spiral [45] broadband antennas to achieve nearly constant antenna impedance in the frequency range above about 100 GHz [53]. Alternatively, bow-tie antennas are used for polarized input and output [52]. In order to suppress back-reflections and for pre-collimation, the terahertz radiation is coupled to free space via a hyper-hemispherical silicon lens. Scanning electron micrograph of a photomixer with integrated spiral antenna [45] is shown in Fig. 3.14a and the SEM picture of an interdigitated electrodes with integrated bow-tie antenna is shown in Fig. 3.14b [52]. Receiver photomixer and the integrated antennas are usually adapted to the transmitter antennas for better sensitivity. The incident THz field is sampled coherently using the same two-color laser as used for the transmitter.

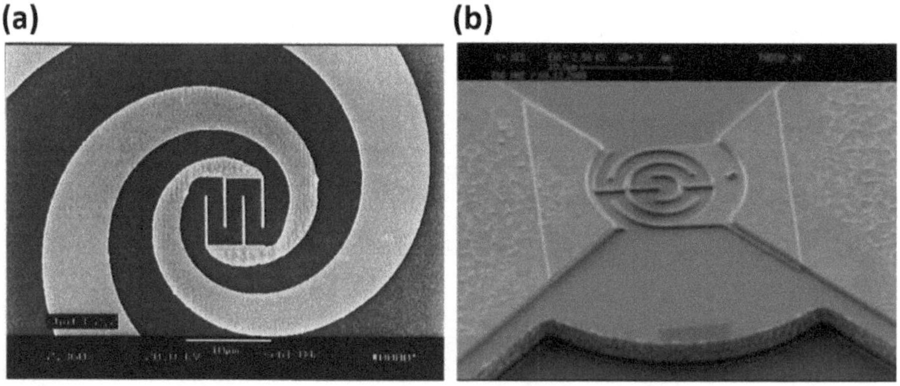

Fig. 3.14 (**a**) Scanning electron micrograph (SEM) picture of electrode region and first turn of spiral antenna for 8 μm × 8 μm photomixer [45]. (**b**) SEM picture of the interdigitated electrodes on the pc material with integrated bow-tie antenna, used in a cw terahertz receiver modules [52]

3.3.2.4 Data Analysis

The detected DC photocurrent (typically $\sim$ pA) is preamplified using a transimpedance amplifier and measured using lock-in technique. The principle of lock-in technique is as described in Sect. 3.3.1. Instead of mechanical chopping, typically, the bias voltage of the transmit photomixer is modulated ($\sim \pm 10$ V and up to 10 kHz), which offers the advantage of less noise and no mechanical parts [47]. The photocurrent I_{ph} is proportional to the incident THz electric field, E_{THz}, and the phase difference $\Delta\phi$ between the THz field and the laser beat frequency [46]

$$I_{\text{ph}} \propto E_{\text{THz}} \cos(\Delta\phi) = E_{\text{THz}} \cos(2\pi \Delta L \nu / c) \tag{3.10}$$

where ν is the terahertz frequency, c is the speed of light, and

$$\Delta L = (L_S + L_{\text{THz}}) - L_D \tag{3.11}$$

is the difference between the optical path L_D traveled by the laser beat to the detector and the optical path L_S of the laser beat to the terahertz source plus the terahertz path L_{THz} from the source to the detector.

One possible way to separate the amplitude and phase of the terahertz signal for any given frequency is to vary L_{THz} and thus ΔL with the help of a variable delay stage [47, 53]. Alternatively, the phase difference $\Delta\phi$ can be varied by scanning the terahertz frequency ν, which has the advantages of avoiding any mechanically moving parts and of a much higher scanning rate (up to 10 Hz, depending on the selected lock-in time). In this case, the detected photocurrent I_{ph} oscillates as a function of frequency (see Fig. 3.15a), and the oscillation period is set by the

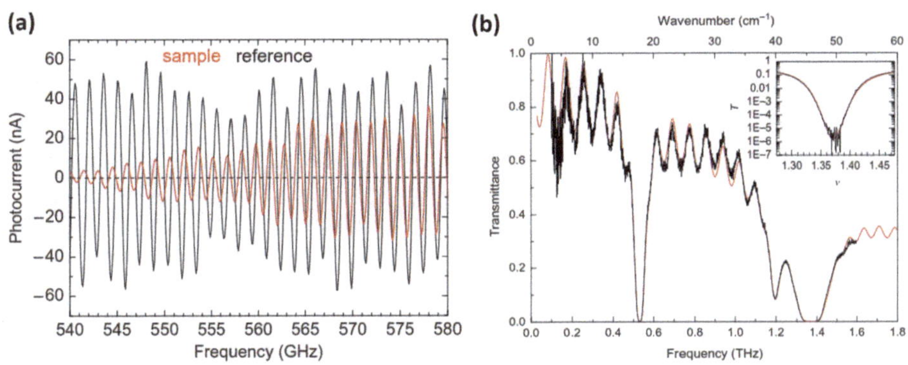

Fig. 3.15 Example of detected terahertz photocurrent and its transmittance plot [46]. (**a**) Photocurrent with and without a sample (α-lactose monohydrate). The period of interference pattern $\Delta\nu \approx 1.5$ GHz with a optical path difference $\Delta L \approx 0.2$ m (see Eq. (3.11)). (**b**) Transmittance $T(\omega)$ of α-lactose monohydrate for a thickness of $d \approx 1$ mm. The *red line* depicts a fit according to the DrudeLorentz model. Inset: $T(\omega)$ around the absorption feature at 1.369 THz, plotted on a log scale

choice of ΔL. Both the amplitude and phase can be derived from it. The resultant spectra is extracted by calculating the envelope of the oscillating photocurrent (see Fig. 3.15b). Using the extremas, the effective frequency resolution for amplitude can be expressed as $c/(2 \times \Delta L)$ and since phase information can be extracted between an extrema and its adjacent zero-crossing, the frequency resolution for phase can be expressed as $c/(4 \times \Delta L)$ [46]. Thus the spectral resolution of the phase can be adapted easily by varying the optical path difference ΔL (see Eq. (3.11)).

The transmittance of a sample $T(\omega)$ can be determined by comparing the envelopes of the sample and reference spectra,

$$T(\omega) = \frac{I_{\mathrm{ph}}^{\mathrm{sample}}(\omega)}{I_{\mathrm{ph}}^{\mathrm{ref}}(\omega)} \tag{3.12}$$

where $I_{\mathrm{ph}}^{\mathrm{ref}}(\omega)$ and $I_{\mathrm{ph}}^{\mathrm{sample}}(\omega)$ are the spectra of the reference and the sample, respectively. For cases where the difference between the sample and reference is small, a useful trick to improve the readability of the data is to plot the differential spectra,

$$\Delta T(\omega) = \frac{I_{\mathrm{ph}}^{\mathrm{sample}}(\omega) - I_{\mathrm{ph}}^{\mathrm{ref}}(\omega)}{I_{\mathrm{ph}}^{\mathrm{ref}}(\omega)} \tag{3.13}$$

An example of the detected terahertz photocurrent with and without a sample (α-lactose monohydrate) and its transmittance $T(\omega)$ is shown in Fig. 3.15 [46].

3.3.3 *Optical Alignment of Off-Axis Parabolic Mirrors*

For every THz spectroscopy system, a set of optics is needed to guide the THz wave from the source to the detector, through the sample if necessary, to make transmissive or reflective measurements (see Sect. 3.3.1.5). An optical system involving reflective surfaces (spherical, paraboloidal, etc.) offers many advantages over refractive systems (Teflon lenses, TPX (poly-4-methyl-pentene-1) lenses, etc). All-reflective optical systems have wide bandwidth and no chromatic aberration, allowing it to be tested and aligned at one frequency (e.g., visible) and spectroscopy at another (e.g., THz) without worrying about realignment.

The parabola is one of the most preferred aspheric surface geometries used for collimating-focusing optical setups. When a collimating beam is incident parallel to the optical axis of the parabola (see Fig. 3.16), the beam is focused to a point image at the focal point of the parabola. Conversely, a point source at the same focal point generates a collimating beam.

In most spectroscopy setups, including THz spectroscopy, the focal point must be placed out of the incident collimated beam [22, 54–56]. For this purpose, the

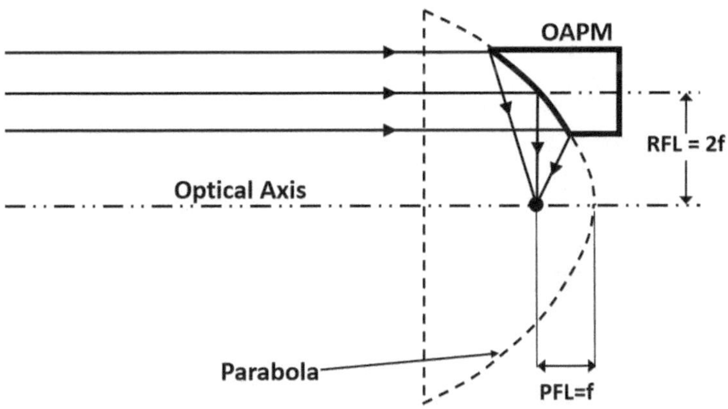

Fig. 3.16 Principle of design of a 90° off-axis parabolic mirror (OAPM) with the reflected focal length (RFL) and the parental focal length (PFL) shown relative to the parabola

most frequently used optics is an off-axis parabolic mirror (OAPM), that is segment cut out of a large parabola as shown in Fig. 3.16. The optical axis of the OAPM is the same as that of the parent parabola which is the axis of revolution.

Given all the advantages of OAPMs, the main disadvantage is difficulty in aligning them. Traditionally, sophisticated interferometer methods such as laser unequal path method [57] or use of alignment telescopes were used to align the OAPMs. This methods were time consuming and prone to large errors. In 1992, Lee published a two step method [58] to align pair of OAPMs using a Helium-Neon (He-Ne) laser, a beam splitter, an iris diaphragm, and a plane mirror. This method makes it possible to align the mirrors without knowing the focal length or the off-axis distance. In the first step, the OAPM is adjusted such that the optical axis is in plane with two incident parallel beams. In the next step, it is aligned such that the optical axis is in plane of the incident beams, in the direction pointed by the incident beams. In this section we will describe an alignment method based on Lee's two step method [58], but replacing the He-Ne laser, the plane mirrors and the iris with simple laser diodes and alignment scales.

3.3.3.1 Alignment Procedure

The alignment procedure is as described in Lee's work in [58] but adapted to use collimated laser diodes and fixed apertures instead of the He-Ne laser with beam spliters and plane mirrors. Figure 3.17 shows the basic arrangement of the laser diodes (LD1, LD2) and off-axis parabolic mirror (OAPM) to be aligned. A typical 635 nm collimated laser diode modules with a round beam size of ~2–3 mm can be used for the laser sources. For alignment, the lasers can be mounted on a Kinematic Mount. The two-laser beams (Beam 1 and Beam 2) are made parallel to the surface of the optical table using a set of fixed apertures (~2–3 mm), that are set at the same

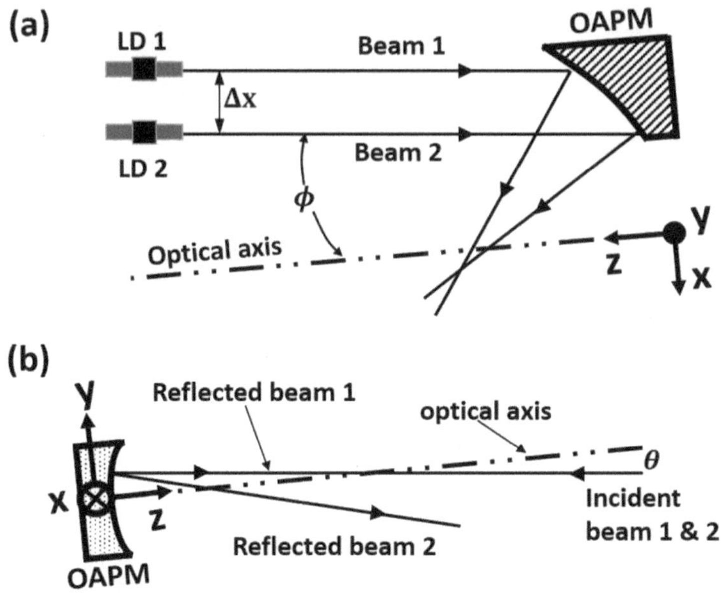

Fig. 3.17 (**a**) *Top view* and, (**b**) *side view* of the optical table for vertical alignment

height. Alternatively, Magnetic Beam Height Measurement Tool with alignment holes can be used to make the lasers parallel to the optical table surface. The laser beams are also made parallel to each other. If using fixed apertures mounted on posts in the optical slots, the beams should automatically get parallel to themselves as well. The distance between the beams Δx is determined by the width of the OAPM. It can be shown that the optical axis of OAPM get aligned better with the incident beams for larger Δx [58].

In this section we use the same rectilinear coordinate system as used in Lee's method [58], as shown in Fig. 3.17. In the first alignment, the vertical alignment, the optical axis of the OAPM is placed in the plane created by the two incident beams. In the next alignment, the horizontal alignment, the optical axis is made parallel to the incident beams. The focal point of the OPAM is found where the two reflected beams meet.

3.3.3.2 Vertical Alignment

In order to bring the optical axis in plane of the incident beams, the steps for the vertical alignment procedure are:

1. The OAPM is placed on the optical table using kinematic mounts and the optical axis is made approximately parallel to the incident beams as shown in Fig. 3.17a.

2. Then the reflected beam 1 is made parallel to the optical table. The same fixed apertures can be used at two different locations to check the parallelism. A Magnetic Beam Height Measurement Tool with alignment holes can be used as well.
3. Next we make the reflected beam 2 parallel to the optical table using the same method as used above for the reflected beam 1.
4. If the reflected 2 beam moves upward, then we decrease the height of the OAPM.
5. Similarly, if the reflected beam 2 moves downward, we increase the height of the OAPM (see Fig. 3.17b).

Steps 2–5 are repeated till the both the reflected beams are parallel to the surface of the optical table, thus bringing the optical to the plane of the incident beams.

3.3.3.3 Horizontal Alignment

Next, in order to make the optical axis parallel to incident beams, the steps for the horizontal alignment procedure are:

1. A plane mirror, preferably mounted on a kinematic mount, is placed at the intersection of the two reflected beams. This is not necessarily the focal point since the optical axis may not be parallel to the incident beams yet.
2. The plane mirror is then tilted to send one of the beams to a point on the OAPM such that the reflected beam (beam 3 in Fig. 3.18a, b) is above and in-between the incident beams. To make sure the plane mirror is at the meeting point of the two beams, a thin paper can be held at the surface of the mirror to see the convergence of the two beams.
3. Next, the reflected beam 3 is checked to see if it is parallel to the optical table. It can be done by moving an iris or a fixed aperture on the table with a fixed height. And like before, a Magnetic Beam Height Measurement Tool with alignment holes can be used as well.
4. If the reflected beam 3 is propagating upwards, we need to rotate the OAPM about the y-axis in the direction of the smaller angle of incidence of beam 1 and 2.
5. Conversely, if the reflected beam 3 is propagating downwards, we need to rotate the OAPM about the y-axis in the direction of the larger angle of incidence of beam 1 and 2.

Steps 1–5 are repeated till the reflected beam is parallel to the optical table. Then the optical axis becomes parallel to the incident beams 1 and 2 and the focal point of the OAPM is where the reflected beam 1 and 2 meet.

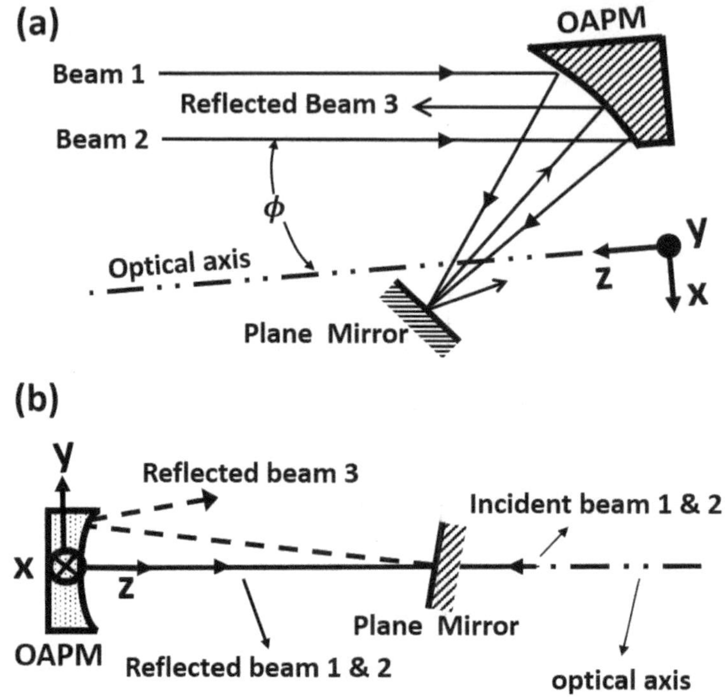

Fig. 3.18 (**a**) *Top view* of optical table for horizontal alignment. A plane mirror is put where the reflected beam 1 and 2 cross each other. (**b**) *Side view* of optical table for horizontal alignment. An angular deviation of the reflected beam 3 is related to a rotation of the optical axis about y-axis

References

1. D.R. Smith, S. Schultz, P. Marko, C.M. Soukoulis, Phys. Rev. B **65**(19), 195104 (2002)
2. C.G. Parazzoli, R.B. Greegor, K. Li, B.E.C. Koltenbah, M. Tanielian, Phys. Rev. Lett. **90**(10), 107401 (2003)
3. X. Chen, T.M. Grzegorczyk, B.I. Wu, J. Pacheco, J.A. Kong, Phys. Rev. E **70**(1), 016608 (2004)
4. D.R. Smith, D.C. Vier, T. Koschny, C.M. Soukoulis, Phys. Rev. E **71**(3), 036617 (2005)
5. X. Chen, B.I. Wu, J.A. Kong, T.M. Grzegorczyk, Phys. Rev. E **71**(4), 046610 (2005)
6. MICROWAVE STUDIO is the trademark of CST GmbH, Darmstadt, Germany (2014)
7. J.B. Pendry, A.J. Holden, W.J. Stewart, I. Youngs, Phys. Rev. Lett. **76**(25), 4773 (1996)
8. V.M. Shalaev, W. Cai, U.K. Chettiar, H.K. Yuan, A.K. Sarychev, V.P. Drachev, A.V. Kildishev, Opt. Lett. **30**(24), 3356 (2005)
9. W.J. Padilla, D.N. Basov, D.R. Smith, Mater. Today **9**(7–8), 28 (2006)
10. D. Mittleman, *Sensing with Terahertz Radiation* (Springer, Berlin, 2013)
11. M. Tonouchi, Nat. Photonics **1**(2), 97 (2007)
12. W. Wu, E. Kim, E. Ponizovskaya, Y. Liu, Z. Yu, N. Fang, Y.R. Shen, A.M. Bratkovsky, W. Tong, C. Sun, X. Zhang, S.Y. Wang, R.S. Williams, Appl. Phys. A **87**(2), 143 (2007)
13. N. Dutta, S. Shi, D. Prather, in *CLEO:2011 - Laser Applications to Photonic Applications (2011), paper CMV3* (Optical Society of America, 2011), p. CMV3
14. S.Y. Chou, P.R. Krauss, P.J. Renstrom, Science **272**(5258), 85 (1996)

15. J.D. Albrecht, M.J. Rosker, H.B. Wallace, T.H. Chang, in *2010 IEEE MTT-S International Microwave Symposium Digest (MTT)* (2010), pp. 1118–1121
16. B.E.A. Saleh, M.C. Teich, *Fundamentals of Photonics* (Wiley, New York, 2007)
17. N.W. Ashcroft, N.D. Mermin, *Solid State Physics* (Cengage Learning, New Delhi, 2011)
18. M.A. Ordal, L.L. Long, R.J. Bell, S.E. Bell, R.R. Bell, R.W. Alexander, C.A. Ward, Appl. Opt. **22**(7), 1099 (1983)
19. D. Schurig, J. Mock, B. Justice, S. Cummer, J. Pendry, A. Starr, D. Smith, Science **314**(5801), 977 (2006)
20. R. Liu, A. Degiron, J.J. Mock, D.R. Smith, Appl. Phys. Lett. **90**(26), 263504 (2007)
21. Y. Yuan, C. Bingham, T. Tyler, S. Palit, T.H. Hand, W.J. Padilla, D.R. Smith, N.M. Jokerst, S.A. Cummer, Opt. Express **16**(13), 9746 (2008)
22. D. Shrekenhamer, S. Rout, A.C. Strikwerda, C. Bingham, R.D. Averitt, S. Sonkusale, W.J. Padilla, Opt. Express **19**(10), 9968 (2011)
23. D.H. Auston, A.M. Glass, P. LeFur, Appl. Phys. Lett. **23**(1), 47 (1973)
24. D.H. Auston, K.P. Cheung, JOSA B **2**(4), 606 (1985)
25. A.P. DeFonzo, M. Jarwala, C. Lutz, Appl. Phys. Lett. **50**(17), 1155 (1987)
26. P.R. Smith, D.H. Auston, M.C. Nuss, IEEE J. Quantum Electron. **24**(2), 255 (1988)
27. C. Fattinger, D. Grischkowsky, Appl. Phys. Lett. **54**(6), 490 (1989)
28. D. Grischkowsky, S. Keiding, M.v. Exter, C. Fattinger, J. Opt. Soc. Am. B **7**(10), 2006 (1990)
29. B.B. Hu, M.C. Nuss, Opt. Lett. **20**(16), 1716 (1995)
30. D.E. Spence, P.N. Kean, W. Sibbett, Opt. Lett. **16**(1), 42 (1991)
31. R.L. Fork, B.I. Greene, C.V. Shank, Appl. Phys. Lett. **38**(9), 671 (1981)
32. S. Tsuda, W.H. Knox, E.A.d. Souza, W.Y. Jan, J.E. Cunningham, Opt. Lett. **20**(12), 1406 (1995)
33. D. Auston, IEEE J. Quantum Electron. **19**(4), 639 (1983)
34. D.H. Auston, K.P. Cheung, P.R. Smith, Appl. Phys. Lett. **45**(3), 284 (1984)
35. N. Katzenellenbogen, D. Grischkowsky, Appl. Phys. Lett. **58**(3), 222 (1991)
36. M. van Exter, D.R. Grischkowsky, IEEE Trans. Microwave Theory Tech. **38**(11), 1684 (1990)
37. S. Gupta, M.Y. Frankel, J.A. Valdmanis, J.F. Whitaker, G.A. Mourou, F.W. Smith, A.R. Calawa, Appl. Phys. Lett. **59**(25), 3276 (1991)
38. F.E. Doany, D. Grischkowsky, C.C. Chi, Appl. Phys. Lett. **50**(8), 460 (1987)
39. M.C. Nuss, J. Orenstein, in *Millimeter and Submillimeter Wave Spectroscopy of Solids*, ed. by P.D.G. Grner. Topics in Applied Physics, vol. 74 (Springer, Berlin/Heidelberg, 1998), pp. 7–50
40. G.W. Chantry, H.M. Evans, J.W. Fleming, H.A. Gebbie, Infrared Phys. **9**(1), 31 (1969)
41. K.G. Libbrecht, E.D. Black, C.M. Hirata, Am. J. Phys. **71**(11), 1208 (2003)
42. J.H. Scofield, Am. J. Phys. **62**(2), 129 (1994)
43. M.L. Meade, J. Phys. E: Sci. Instrum. **15**(4), 395 (1982)
44. L. Ning, S. Jingling, S. Jinhai, L. Laishun, X. Xiaoyu, L. Meihong, J. Yan, Opt. Express **13**(18), 6750 (2005)
45. K.A. McIntosh, E.R. Brown, K.B. Nichols, O.B. McMahon, W.F. DiNatale, T.M. Lyszczarz, Appl. Phys. Lett. **67**(26), 3844 (1995)
46. A. Roggenbuck, H. Schmitz, A. Deninger, I.C. Mayorga, J. Hemberger, R. Gusten, M. Gruninger, New J. Phys. **12**(4), 043017 (2010)
47. A.J. Deninger, T. Gbel, D. Schnherr, T. Kinder, A. Roggenbuck, M. Kåberle, F. Lison, T. Mller-Wirts, P. Meissner, Rev. Sci. Instrum. **79**(4), 044702 (2008).
48. M. McGehee, M. Diaz-Garcia, R. Gupta, F. Hide, A.J. Heeger, in *IEEE Lasers and Electro-Optics Society Annual Meeting, 1997. LEOS '97 10th Annual Meeting. Conference Proceedings*, vol. 2 (1997), pp. 172–174
49. I.S. Gregory, M.J. Evans, H. Page, S. Malik, I. Farrer, H.E. Beere, Appl. Phys. Lett. **91**(15), 154103 (2007)
50. M. Scheller, M. Stecher, M. Gerhard, M. Koch, Opt. Express **18**(15), 15887 (2010)
51. S. Kraft, A. Deninger, C. Trck, J. Fortgh, F. Lison, C. Zimmermann, Laser Phys. Lett. **2**(2), 71 (2004)

52. D. Stanze, A. Deninger, A. Roggenbuck, S. Schindler, M. Schlak, B. Sartorius, J. Infrared Millimeter Terahertz Waves **32**(2), 225 (2010)
53. R. Mendis, C. Sydlo, J. Sigmu nd, M. Feiginov, P. Meissner, H.L. Hartnagel, IEEE Antennas Wirel. Propag. Lett. **4**, 85 (2005)
54. C.M. Watts, D. Shrekenhamer, J. Montoya, G. Lipworth, J. Hunt, T. Sleasman, S. Krishna, D.R. Smith, W.J. Padilla, Nat. Photonics **8**, 605 (2014)
55. S. Rout, S. Sonkusale, Opt. Express **24**(13), 14618 (2016)
56. S. Rout, S.R. Sonkusale, APL Photon. **1**(8), 086102 (2016)
57. C. Bond, C.A. Pipan, How to align an off-axis parabolic mirror, in *Proceedings of SPIE*, vol. 1113. doi:10.1117/12.955592; Reflective Optics II, 236 (October 11, 1989)
58. Y.H. Lee, Opt. Eng. **31**(11), 2287 (1992)

Chapter 4
High-Speed Terahertz Modulation Using Active Metamaterial

This chapter covers a case study of metamaterial based terahertz modulator using embedded HEMT devices. A computational and experimental study of the terahertz (THz) device resulting from hybridization of metamaterials with pseudomorphic high electron mobility transistors (pHEMTs), fabricated in a commercial gallium arsenide (GaAs) process is presented. The principle of modulation introduced in Sect. 2.3 is further elaborated with a detailed analysis of HEMT based modulation. The design and fabrication details of the device using a commercial Gallium-Arsenide (GaAs) process are covered next and finally the experimental methods and the results are discussed.

4.1 Introduction

Electronically controlled high-speed terahertz (THz) modulation is highly desirable due to its vast potential of creating low-cost consumer grade devices for applications ranging from communication to imaging. The demand for high-speed wireless access is increasing due to large amounts of data needed for new emerging applications such as the consumer market that is already demanding 20, 40, and 100 Gbit/s wireless technologies for Super Hi-Vision (SHV) and Ultra High-Def (UHD) TV data [1]. Terahertz (THz) carrier frequencies will offer the advantage of higher data speed, sub-millimeter antenna size and short range security especially suitable for portable devices. Although they are susceptible to atmospheric loss, THz digital communication systems have been demonstrated near certain windows, especially around the 300–400 GHz range that has shown promise for high bit-rate data transmission [1–3]. In addition to THz sources and detectors, one of the key components of a THz communication system is a modulator that is used to modulate the carrier waves with data streams. Designing high-speed modulators

S. Rout, S. Sonkusale, *Active Metamaterials*, DOI 10.1007/978-3-319-52219-7_4

for the frequency regime (0.3–10 THz, $\lambda = 1$ mm–30 μm), so-called terahertz gap [4], has been a difficult primarily due to lack of suitable materials for constructing electronic or quasi-optical devices.

In recent years, various methods have been utilized to achieve amplitude and frequency modulation including photodoping [5], electronic [6], and temperature control [7]. By implementing schottky diodes other research has achieved both spatial [8, 9] and phase modulation [10]. These metamaterial devices rely on layers of n-doped gallium arsenide (GaAs) and initial studies have demonstrated the potential of these systems for high-speed telecommunications [11], spectroscopy [10], and imaging [12]. To-date THz metamaterials have shown modulation at frequencies low in the MHz range, and higher speed modulation has been hindered by large device capacitance [8], making them unsuitable for the applications that are demanding high-speed and low-voltage operations.

The motivation to create high-speed modulators has pushed researchers to use two-dimensional electron gases (2DEGs) in high electron mobility transistors (HEMTs) to manipulate THz waves [13]. Although HEMT devices have demonstrated advantages over alternative architecture, such as quantum well structures and liquid crystals [14], they have been limited to relatively low transmission modulation values of less than 3%, and maximum modulation frequencies of 10 kHz [15]. One of the keys to high-speed terahertz devices is the hybridization of semiconductor technology and metamaterials, an area of research still in its infancy.

In this chapter, a metamaterial based THz modulator with embedded HEMT has been demonstrated to operate at high speed (∼10 MHz) and operating at low voltage (1 V) makes it suitable for high-speed and low-power communication [16] and imaging systems [17]. The results illustrate modulation of terahertz frequencies (0.45 THz) up to speeds of 10 MHz, which can fundamentally be designed for much higher speeds.

4.2 Design Principle of the HEMT Controlled Metamaterial Modulator

The basic principle of our THz SLM is based on an electronically controlled absorption coefficient in a metamaterial using embedded pHEMT devices as described below. For a metamaterial designed with electric-LC (ELC) resonators [18, 19] shown in Fig. 4.1a, the average permittivity, without factoring spatial dispersion, can be expressed in a Drude-Lorentz form [20, 21]

$$\bar{\epsilon}(f) = \epsilon_a \left[1 - \frac{Ff^2}{f^2 - f_0^2 + i\gamma f}\right] \tag{4.1}$$

where ϵ_a is the permittivity of the base substrate material (silicon nitride and GaAs substrate), f_0 is the resonant frequency in terms of its equivalent circuit parameters [Fig. 4.1d], γ is associated with the loss [R_{LOSS} in Fig. 4.1d], and F is associated with filing factor of the geometry of the unit cell.

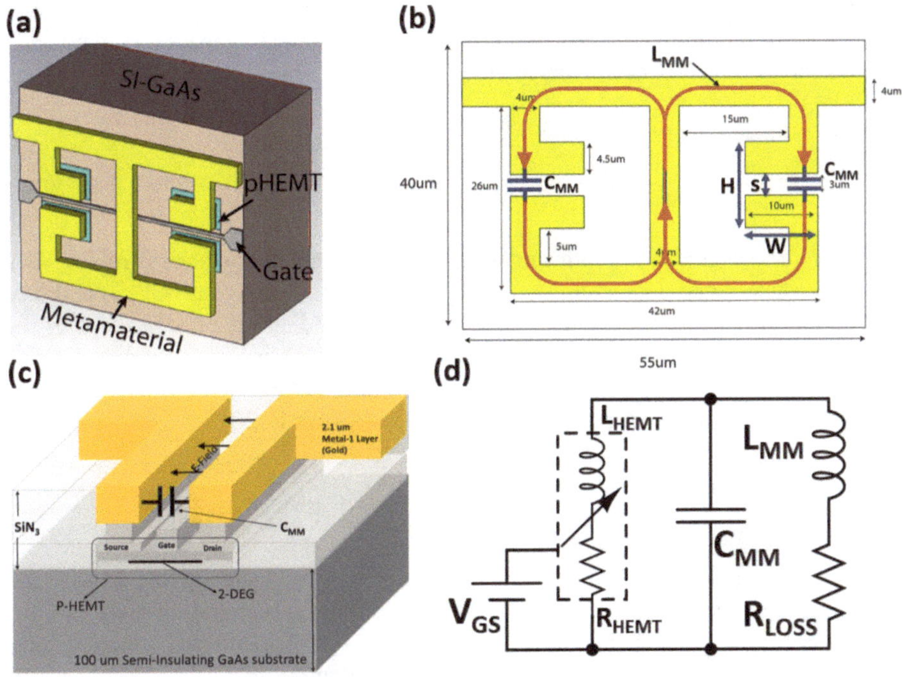

Fig. 4.1 Design and fabrication details. (**a**) Each metamaterial element is based on the electric-LC (ELC) resonator, patterned using the top 2.1 μm thick gold metal. A pseudomorphic high-electron mobility transistor (pHEMT) is placed underneath each split gap with their source and drain connected to each side of the split gap. The gate-to-source/drain (V_{GS}) controls the channel charge (2-DEG) between the split gap, thus electronically controlling the resonance. (**b**) Shows the physical representation of the circuit model L_{MM} and C_{MM}, along with the dimensions of the metamaterial. (**c**) Close-up diagram of the split gap shows the placement of the HEMT device with its drain and source connected to both ends of the split gap. (**d**) An equivalent circuit of the HEMT-embedded metamaterial element, where the resistor-inductor (inside the *dashed box*) represents the Drude model of the HEMT switch at the operating frequency of 0.45 THz when $V_{GS} = 0$ V

4.2.1 Circuit Model for the Electric-Coupled LC (ELC) Resonator

As shown in Fig. 4.1b, the ELC resonator is equivalent to two split-ring resonators (SRRs) put together with a fundamental mode that couples strongly to a uniform electric field linearly polarized perpendicular to the split gap, and negligibly to a uniform magnetic field making it suitable for applications with incident THz wave normal to the planar structure. The ELC resonator can be represented by an equivalent RLC circuit as shown in Fig. 4.1d where in its simplest form, C_{MM} is the capacitor associated with the split gap of the ELC, L_{MM} is the inductor associated with the circulating current in the two symmetric loops as shown in Fig. 4.1a. And R_{LOSS} is the resistive loss of the metal (gold) for the circulating inductor current.

The capacitor C_{MM}, in its simplest form, can be expressed as $C_{MM} = \epsilon_{SiN_3} A/s$ where s and A are the gap width and surface area of the split gap, respectively, and ϵ_{SiN_3} is the permitivity of silicon nitride. A more accurate form can be expressed as [20]

$$C_{MM} = \epsilon_{SiN_3} \frac{2}{\pi} \ln \left(2\alpha \frac{H}{s} \right) W \tag{4.2}$$

where H and W are the height and width of the split gap as shown in Fig. 4.1a and α is a geometry dependent constant. It should be noted that this capacitance calculation still has error in the value due to second order effects such as the periphery and substrate effect. A more accurate expression of this capacitor can be found in [20].

The inductor can be expressed in its first-order form as [20]

$$L_{MM} = \frac{\mu_0 l}{2\pi} \left[ln \left(\frac{2l}{b} \right) + \frac{1}{2} + \frac{b}{3l} + \frac{b^2}{24l^2} \right] \tag{4.3}$$

where l is the length of the inductive loop and b is the width of the conducting strip. This inductor is also a first-order calculation for demonstrating the principle. Second-order effects such as mutual coupling between elements and spatial dispersion needs to be taken account for more accurate calculation.

When the pHEMT is "off" ($V_{GS} = -1$ V), the values for L_{HEMT} and R_{HEMT} in Fig. 4.1d are negligible and the resonant frequency f_0 can be expressed in terms of the equivalent circuit parameters as

$$f_0 = \frac{1}{2\pi \sqrt{L_{MM} C_{MM}}} \tag{4.4}$$

where C_{MM} is the capacitance associated with the split gap of the metamaterial as shown in the close-up drawing in Fig. 4.1c and L_{MM} is the inductance associated with the current loop in each half of the metamaterial loop. The damping factor γ, related to the loss in the ELC, R_{LOSS}, can be expressed as [20]

$$\gamma = \frac{1}{2\pi} \frac{R_{LOSS}}{L_{MM}} \tag{4.5}$$

R_{LOSS} can be approximated as [22]

$$R_{LOSS} = \frac{l}{\sigma P \delta} \tag{4.6}$$

where l is the length of the conductor forming the inductive loop, σ is the conductivity of conductor, P is the perimeter of the cross-section of the conductor, and δ is the skin depth of conductor at f_0.

The ELC usually does not exhibit a strongly dispersive permeability ($\bar{\mu}(f) \approx \mu_0$). Since the dispersion relation is complex, $\bar{\epsilon}(f) = \epsilon_1(f) + i\epsilon_2(f)$, the absorbtion of the transmissive terahertz wave is directly proportional to the imaginary part of the dispersion $\epsilon_2(f)$ and the maximum absorption occurs at the resonant frequency f_0 known as the *anomalous dispersion* [23].

4.2.2 *Principle of Voltage Controlled Terahertz Wave Modulator*

The pHEMT is a standard device offered in the process, that is constructed using pseudomorphic undoped InGaAs and a lightly doped schottky layer, 12.7 nm and 22.5 nm thick, respectively, creating a heterojunction. An equivalent heterojunction is shown in Fig. 4.2a from [24].[1] A very high-mobility 2DEG ($\sim$3000 $\text{cm}^2/\text{V}\cdot\text{s}$) is formed at room temperature in the *undoped* In-GaAs channel layer as predicted by the band diagram at the interface [24], enabling fast conduction even at THz frequencies. The gate voltage with respect to the source (or drain), V_{GS}, controls the charge density in the 2-DEG layer between the split gap. Using the basic layer

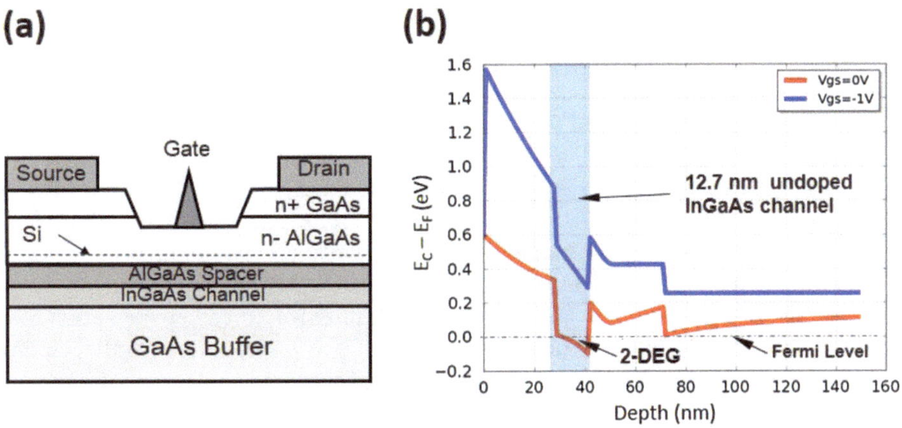

Fig. 4.2 pHEMT heterostructure and energy band diagram. (**a**) Cross-section of a gate-length planar pHEMT from [24] which is similar to the structure used in this work. (**b**) Computed energy-band diagram of the planar doped AlGaAs/InGaAs/AlGaAs quantum well heterostructure of the pHEMT that is used in this work. The band diagram is computed for V_{GS} of 0 V and −1 V. The conduction band discontinuity can be seen in the undoped InGaAs channel. For $V_{GS} = 0$ V the conduction band falls below the fermi level resulting in a high-mobility ($\sim$3000 $\text{cm}^2/\text{V}\cdot\text{s}$) channel with a computed charge density of $N_S = 7.37 \times 10^{12}\text{cm}^{-2}$

[1]Due to reasons of confidentiality with the manufacturing vendor, the proprietary heterostructure could not be published.

information provided by the manufacturer, a 1-D Poisson solver [25] was used to compute the band diagrams for $V_{GS} = 0\,\text{V}$ and $V_{GS} = -1\,\text{V}$ and the results are shown in Fig. 4.2b. The conduction band discontinuity can be seen in the undoped InGaAs channel which is responsible for creating a high-mobility channel for fast conduction. For $V_{GS} = -1\,\text{V}$, the charge carriers in the 2-DEG is completely depleted or the *metamaterial is "on"*. For $V_{GS} = 0\,\text{V}$, the conduction band falls below the fermi level [Fig. 4.2b] resulting in a high-mobility ($\sim 3000\,\text{cm}^2/\text{V}\cdot\text{s}$) channel with a computed charge density of $N_S = 7.37 \times 10^{12}\,\text{cm}^{-2}$ or the *metamaterial is "off"*.

When the pHEMT is "on" ($V_{GS} = 0\,\text{V}$), a high-mobility, two-dimensional electron gas (2-DEG) channel is formed between the drain and the source of the pHEMT which can be described by the Drude conductivity [26]

$$\sigma_{2D}(\omega) = \frac{\sigma_0}{1 + i\omega\tau} \tag{4.7}$$

where σ_0 is the DC conductivity [Siemens/sq] given by the expression $\sigma_0 = e^2 N_S \tau / m^*$. Here, e is the electron charge, N_S is the two-dimensional carrier density of the 2-DEG, τ is the scattering or relaxation time, m^* is the effective electron mass, and ω the angular frequency of the incident time-varying field. The reader can refer to the work by Chao et.al. [24] for a relation between carrier concentration and V_{GS}.

This conductance can be represented as a series R-L structure ($L_{\text{HEMT}}, R_{\text{HEMT}}$) as shown in Fig. 4.1d, that can be expressed as

$$L_{\text{HEMT}} = \frac{\tau}{\sigma_0} \quad \text{and,} \quad R_{\text{HEMT}} = \frac{1}{\sigma_0} \tag{4.8}$$

L_{HEMT} is typically referred to as the *kinetic* inductance [27] and the R_{HEMT} is the DC resistance of the 2-DEG channel along with the resistance of the ohmic layer connecting the pHEMT to the metamaterial split gap as shown in Fig. 4.1c.

In order to find the new resonant frequency f_0' and the damping frequency γ' when the pHEMT is "on" (e.g. $V_{GS} = 0\,\text{V}$), the circuit in Fig. 4.1d can be converted to a single, parallel R-L-C circuit with elements R_E, L_E, C_{MM} using series-parallel circuit transformation [22]. After the transformation, the resonant and damping frequency can be expressed as

$$f_0' = \frac{1}{2\pi\sqrt{L_E C_{\text{MM}}}} \tag{4.9}$$

$$\gamma' = 2\pi (f_0')^2 \frac{L_E}{R_E} \tag{4.10}$$

For the device under consideration, $f_0' \approx f_0$ and γ' is dominated by the pHEMT conductance and therefore, $\gamma' \gg \gamma$.

The principle of modulation can be demonstrated by plotting the imaginary part of the dispersion equation (4.1) which is directly proportional to the absorbtion of

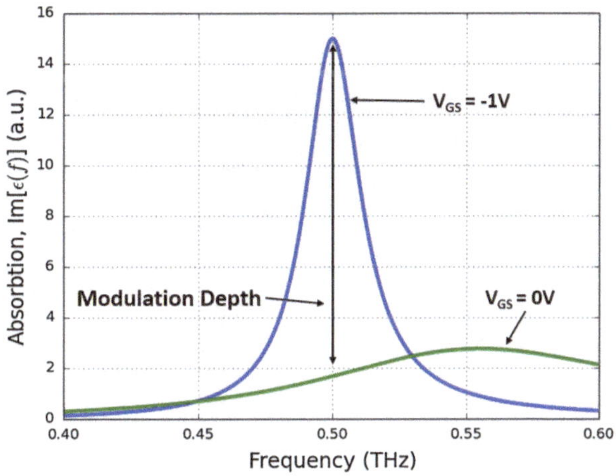

Fig. 4.3 The absorbtion plot [imaginary part of the dispersion equation (4.1)] for $V_{GS} = 0\,\text{V}, f_0 = 0.55\,\text{THz}$, $\Gamma_e = 0.15\,\text{THz}$ (*green*) and $V_{GS} = -1\,\text{V}, f_0 = 0.5\,\text{THz}$, $\Gamma_e = 0.025\,\text{THz}$ (*blue*)

the electromagnetic wave transmitting through the metamaterial sample. For the purpose of demonstration, we choose a some realistic values for $V_{GS} = -1\,\text{V} : f_0 = 0.5$ THz, $\gamma = 0.025$ THz and $V_{GS} = 0\,\text{V} : f_0' = 0.55\,\text{THz}$, $\gamma' = 0.15\,\text{THz}$. Figure 4.3 shows the absorbtion plot for our Drude-Lorentz model dispersion of equation (4.1). As seen from the plot, the primary reason for the modulation is due to the increase in the damping factor or loss in the ELC resonator due to the conductance in the pHEMT that shunts the metamaterial capacitor C_{MM}, thus attenuating the transmissive terahertz wave and hence creating a terahertz modulator at the resonant frequency of the metamaterial.

Because of the localized pHEMT in the split gap, the device has also been demonstrated for fast modulation [16] (~10 MHz) that is capable of much higher speed with proper design. Compared to other electrically controlled metamaterial [6, 10], where the whole substrate is used to control the resonance, this device offers element-level control that offers higher switching speed and offers future opportunity of creating more exotic devices. One such exotic device platform is to populate such amplitude modulator in an array to implement a terahertz spatial light modulator (SLM) without any moving parts. A metamaterial based SLM also offers the advantage of creating hyper-spectral imaging by creating arrays with different frequencies [28].

4.3 Design and Fabrication

The metamaterial design was fabricated using a commercial GaAs process with an active device that is a planar-doped pseudomorphic HEMT (pHEMT) based on AlGaAs-InGaAs-AlGaAs quantum well heterostructure [24]. The technology utilized for this study is primarily used for mobile phone applications and the

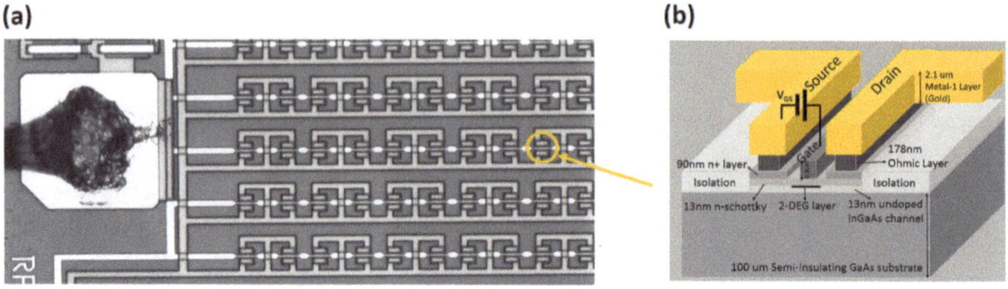

Fig. 4.4 (**a**) A close-up microscope photograph of the metamaterial array near the pad. All the gates of the HEMTs can be seen to be connected together to the pad shown in the picture. Similarly the drain and sources are connected together to another pad (not in the picture). (**b**) 3-D drawing of the cross-section of one of the split gaps with HEMT connected to the metamaterial split gap

THz metamaterial design was adapted for this technology. Based on the design principle explained in the previous section, the metamaterial device is constructed of a planar array of sub-wavelength-sized ELC resonators [18, 19] using the top metal layer, 2.1 μm thick gold (Fig. 4.1a). Each unit cell consists of two single rings butted together with their split gaps at the outside to accommodate design rules specific to the implemented technology. A different metal layer forms the gate of the device and is also used for connecting all the gates within the same row as shown in Fig. 4.4a. A 0.176 μm thick ohmic layer, which is also utilized as a routing layer, was primarily used for connecting the source and drain of the HEMT to the metamaterial (Fig. 4.1c). The line width of the metamaterial is 4 μm and the split gap is 3 μm. The metamaterial had the dimensions of 42 μm wide by 30 μm in height. A periodic array of these unit cells as shown in Fig. 4.4a was fabricated, with period of 55 μm × 40 μm, and a total size of $2.0 \times 2.6\,\text{mm}^2$ with 2200 elements total. Metamaterial elements were fabricated on a 100 μm thick semi-insulating (SI) GaAs substrate.

A pHEMT lies underneath each of the split gaps of the metamaterial element, (two per unit cell), as shown in cross-section in Fig. 4.4b. The gate length is 0.5 μm and has a width 5 μm for each device. The HEMT is constructed using pseudomorphic undoped In-GaAs and a lightly doped Schottky layer, each 12.7 nm thick, creating a heterojunction. A 2DEG is formed in the undoped In-GaAs channel layer as predicted by the band diagram at the interface (Fig. 4.2b) [24]. Unlike traditional FETs, this channel is formed in an intrinsic (undoped) crystal, resulting in very high mobility ($\sim 3000\,\text{cm}^2/\text{V} \cdot \text{s}$) and charge density ($7.37 \times 10^{12}\,\text{cm}^{-2}$) at room temperature, thus enabling fast conduction even at THz frequencies.

The gate voltage with respect to the source (or drain), V_{GS}, controls the charge density in the 2-DEG layer between the split gap and thus changing the resonant frequency as explained in the previous section. When the metamaterial is "on" ($V_{GS} = -1$ V), the charge carriers in the 2-DEG is completely depleted resulting in the metamaterial functioning at its designed resonant frequency (f_0). And when it is "off" ($V_{GS} = 0$ V), the high-mobility channel is formed between the split gap thus shifting the resonant frequency (f_0') as explained in the previous section.

The same metal layer which is used to form each metamaterial is also used to connect each element together within the same row as shown in Fig. 4.4. These wires run perpendicular to the split gaps, and we polarize the electric field of incident radiation perpendicular to the connecting wires which avoids the Drude-like response documented in prior works [6]. At the perimeter of the device each row is connected vertically using the ohmic layer and all elements are connected to a single bond pad to provide DC bias voltage for the drain and source of the HEMT. The gates for all HEMTs are connected in a similar fashion to a single bond pad which provides the DC bias voltage for the gate (see Fig. 4.4).

The entire unit cell was modeled using a commercial finite difference time-domain (FDTD) solver, CST's Microwave Studio, in which the metamaterial was designed to be resonant at 0.46 THz. The physical dimensions of the material layers in the device were modeled as shown in Fig. 4.1b. The gold and ohmic layers were modeled as lossy metals based on their respective DC conductivity values. For both the n+ and Schottky semiconductor layers, we utilize a frequency dependent Drude model for the conductivity. The 2DEG was simulated as a 2 nm thick Drude layer, which enables an accurate modeling of the HEMT device in both the conductive ("off" state) and the depleted ("on" state). This is representative of what occurs in experiment by applying gate to source voltage (V_{GS}) of 0 V (conductive) and −1 V (depleted). This method allowed us to simulate the THz transmission of the device for various DC biases.

4.4 Experimental Setup

A terahertz time-domain spectroscopy (THz-TDS), shown in Fig. 4.5, is used to demonstrate and characterize the HEMT embedded metamaterial (MM) based terahertz (THz) modulator. This setup is used to characterize the metamaterial for DC-biased HEMTs. A detailed description of THz-TDS system is covered in Sect. 3.3.1.

A 532 nm wavelength 4 W neodymium-doped yttrium aluminum garnet (YAG) laser beam is used to pump the titanium sapphire (Ti-sapphire) ultrafast laser that typically outputs a train of optical pulses centered about 800 nm wavelength (∼100 fs in duration), at a repetition rate that varies from 70 to 100 MHz. This ultrafast pulsed laser is split into two different paths using a beam splitter. One of the paths is focused onto a photoconductive substrate gap between the two antenna arms fabricated on a GaAs substrate (Band gap of 1.42 eV = 874 nm). This fast (70–100 MHz) periodic excitation causes generation of carriers which are accelerated towards the antenna due to DC electric fields (typically DC bias voltage of 16 V). The transit time of these carriers in high electric field causes picosecond transient currents. These currents are made to radiate through antenna like structure on either side of electrode, which radiates THz electromagnetic radiation where the central emission frequency can be controlled by virtue of the antenna arms spacing (∼0.3 THz for 80 μm spacing), and $E_{THz}(t)$ is proportional to the time derivative of

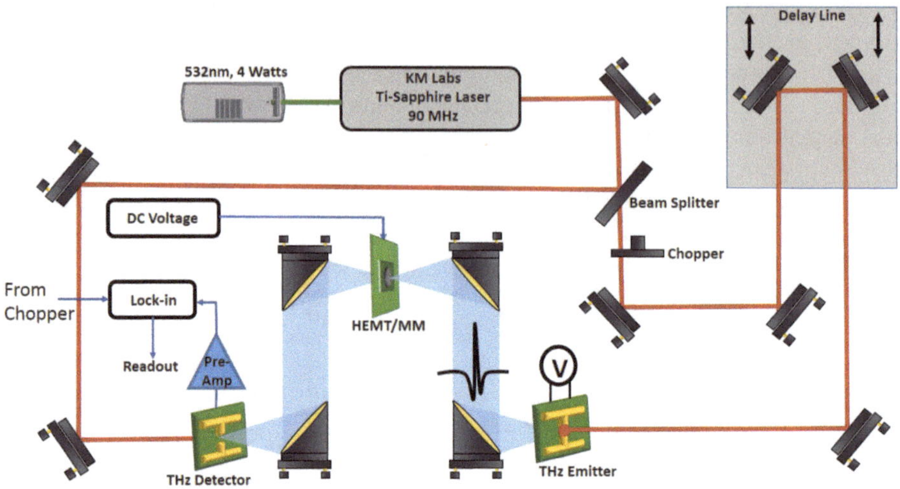

Fig. 4.5 Schematic of the terahertz time-domain spectroscopy (THz-TDS) setup used for characterization of the DC-biased controlled metamaterial based terahertz modulator

the current density generated. This radiated broadband THz wave is collimated, to be carried over a long distance, and focused through the sample, recollimated and focused on to the detector.

The second path of the split pulsed laser is focused onto the photoconductive substrate gap of the detector through a controllable delay stage. The detector is sensitive to the incident THz electric field $E_{\mathrm{THz}}(t)$ for time increments that are much less than 1 ps, thus by scanning the delay line a measure of the $E_{\mathrm{THz}}(t)$ as a function of time can be accomplished where the Fourier transform of the time signal calculates the frequency dependent complex electric field spectrum $E_{\mathrm{THz}}(f)$. The spectral resolution is inversely proportional to the length of the temporal scan.

On the detector side, a dipole antenna is used on top of a substrate material with optically generated carriers that boast very short carrier lifetimes, such as low temperature grown LT-GaAs, Er:GaAs nanoislands, or ion-implanted silicon-on-sapphire (SOS) [29]. The antenna is unbiased and a current is generated by combination of the photogenerated carriers which are accelerated by local THz electric field vector. Due to the short carrier lifetime, the photocurrent can be written as [30]

$$I(t) = \frac{1}{T}\int_0^T E_{\mathrm{THz}}(\tau)\sigma(t+\tau)d\tau \tag{4.11}$$

where T is the repetition rate of the laser source, $\sigma(t)$ is the time dependent conductivity relating the free carriers optically excited in the substrate. If $\sigma(t)$ is a delta function the photocurrent will represent directly E_{THz}. This current is passed to a pre-amplifier and lock-in amplifier and outputs a measured DC voltage proportional to $E_{THz}(t)$. The reference signal for the lock-in signal comes from the optical chopper for coherent detection of the DC signal and remove any DC offset or low-frequency noise in the system as explained in Sect. 3.3.1.

4.5 Results and Discussion

4.5.1 THz Transmission with DC-Biased HEMT

The device was characterized using a THz-TDS as discussed in the previous section. The incident time-domain THz electric field ($E_i(t)$) was polarized along the split gap to drive the metamaterial elements into resonance. At the resonant frequency of the metamaterial, (0.46 THz), the electric field is concentrated within the split gaps of the metamaterial and is within the linear regime of the pHEMT device owing to the low incident power used in experiments. The reference signal was measured with the sample removed from the beam path in order to normalize the data and calculate the absolute transmission. In Fig. 4.6a we show the transmitted electric field as a function of frequency for different V_{GS} values. For V_{GS} less than -1.0 V, the channel is completely depleted, and transmission shows a resonance at 0.46 THz. When the gate-to-source voltage is increased above -1.0 V, the channel starts forming between the split gaps, and the metamaterial resonance begins to diminish. At $V_{GS} = 0$ V, when the channel is completely formed, shunting the metamaterial split gap capacitor C_{MM} (see Fig. 4.1d), effectively weakening the metamaterial resonant response and hence making the metamaterial ineffective. It can be seen in the transmission data (Fig. 4.6a) that the frequency response shows no resonance at $V_{GS} = 0$ V. The principle behind the THz modulation has been detailed in Sect. 4.2. In order to emphasize the switching ability of the terahertz metamaterial, we plot the differential transmission in Fig. 4.6b, defined as

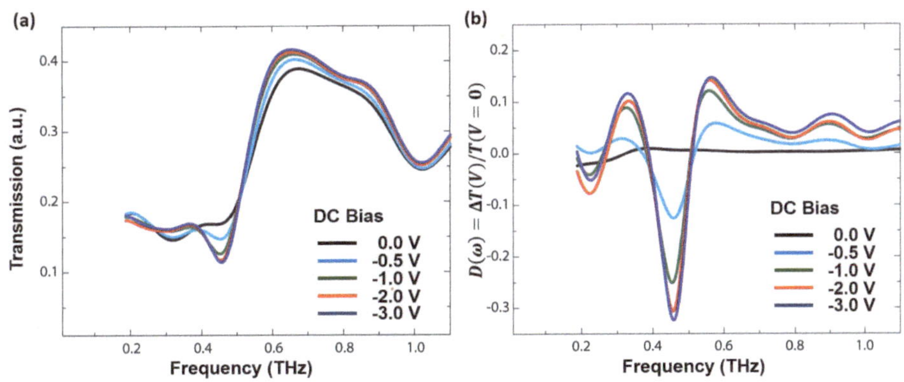

Fig. 4.6 (**a**) Transmittance spectrum of the electric field as function of bias. (**b**) Differential transmission of the electric field as defined in Eq. (4.12)

$$\Delta T_{V_{GS}}(\omega) = \frac{T_{V_{GS}}(\omega) - T_{0V}(\omega)}{T_{0\,V}(\omega)} \tag{4.12}$$

The black curve of Fig. 4.6 is a so-called 100% line, and is two successive transmission measurements divided by each other, both at $V_{GS} = 0$. For a differential transmission of $V_{GS} = -0.5$ V, cyan curve of Fig. 4.6b, $\Delta T_{-0.5\,V}(\omega)$ is relatively flat with deviations from 100% of about 10% or less. This minimum in differential transmission is seen to increase for increasing V_{GS}, until at $V_{GS} = -3.0$ where a value of $\Delta T(\omega) \approx 33\%$ is observed, again at 0.46 THz.

4.5.2 Computational Investigation

In recent years there has been considerable interest in 2-DEG formation in HEMTs as a potential candidate for far-field THz wave interaction [13, 14, 31–34]. A majority of these applications revolve around utilizing the 2DEG instability in the HEMT channel. In this device the plasmon resonance resulting from the 2-DEG instability does not play a roll. As such the parameter of interest for us is the near-field frequency dependent conductivity $\sigma(\omega)$ of the 2-DEG. The dynamic response of the channel carriers can in many cases be described by the two-dimensional Drude conductivity (Eq. (4.7)) [26, 27]. It was also shown in Eq. (4.8) that the real and imaginary part of the conductivity can be represented as a series inductor and resistor in an equivalent circuit model as shown in Fig. 4.1d. It can be observed from Eq. (4.7) that for low frequencies, the conductivity is purely real and hence the current is in phase with the electric field in the split gap, for frequencies near the scattering time-constant both the real and imaginary part is half the DC conductivity and the phase is 45°. For much higher frequencies the conductivity asymptotically approaches zero.

Modification of charge density in a true 2-DEG has no implications on the capacitance of a metamaterial owing to its infinitesimal thickness. However, the nature of our FDTD is inherently three-dimensional. Thus, if we simply change the carrier density, this would falsely modify the capacitance of the metamaterial. In order to approximate the two-dimensional charge density we restrict the real part of the permittivity to be equal to epsilon infinity (12.9 ϵ_0 for GaAs) and allow the imaginary part to be modeled by the Drude model. The resulting complex permittivity can be expressed analytically as

$$\epsilon(\omega) = \epsilon_\infty + i\omega_p^2 \frac{\omega_c/\omega}{\omega^2 + \omega_c^2} \tag{4.13}$$

where ω_c is the collision frequency and ω_p is the plasma frequency. The collision frequency $\omega_c = 2\pi \times 0.7$ THz is calculated with the relation $\omega_c = e/m^*\mu$ where μ is the mobility of the channel along with both e and m^* the electron charge and effective mass in GaAs. The plasma frequency relation is $\omega_p^2 = e^2 N_s \,/\, \epsilon_0 m^* d$

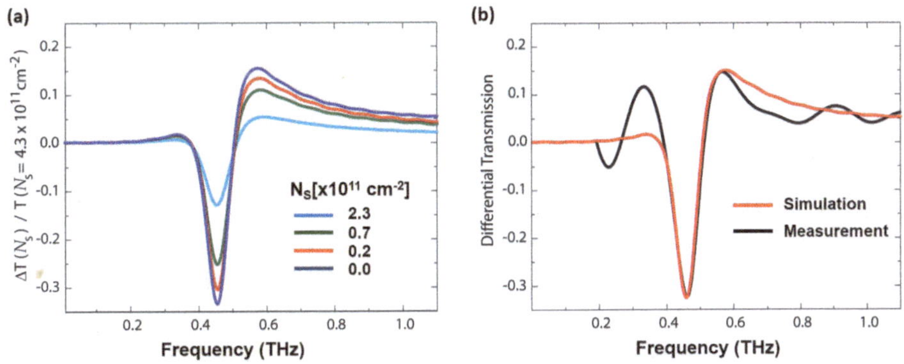

Fig. 4.7 (**a**) Differential transmission simulated for various 2-DEG carrier concentrations. (**b**) A comparison of simulated and experimental differential transmission

where N_s is the two-dimensional carrier concentration and d is the simulated model thickness of the channel layer. We sweep the carrier concentration (See Fig. 4.7a) in simulation observing a full resonant metamaterial response with a completely depleted channel ($N_s = 0\,\text{cm}^{-2}$) and a shorted response as the carrier concentration increased to ($N_s = 7.3 \times 10^{12}\,\text{cm}^{-2}$). Comparing this simulation with the experimental data shows excellent agreement (See Fig. 4.7b).

4.5.3 *High Frequency THz Modulation*

In Sect. 4.5.1 we demonstrated the ability to modulate the THz waveform by adjusting the gate bias voltage of the HEMT with respect to the drain and source (V_{GS}). We now turn toward demonstration of high-speed dynamic modulation. The experimental setup used in Sect. 4.5.1 (Fig. 4.5) is only capable of measuring steady state signal in the system. In order to measure the dynamic THz modulation, the standard mechanical chopper used in system was replaced with our HEMT/metamaterial modulator which serves the same function as the mechanical chopper with the important distinction that now only a narrow band of frequencies about the metamaterial resonance is modulated. The modified experimental setup is shown in Fig. 4.8. Since the effective chopping is only for one frequency (~0.46 THz), a complete scan will result in a sine wave of that frequency, which is essentially the Fourier component of the THz pulse at 0.46 THz. Therefore, this is in indirect but very powerful method to validate the real-time modulation of THz wave.

In this experiment, a square-wave bias is applied to the gate of the HEMT with respect to the source and drain, alternating between -1.1 V and 0 V. The same square wave signal was applied to the reference input of the lock-in amplifier. After collecting the entire time-domain THz signal we have a sampled signal which can be expressed as:

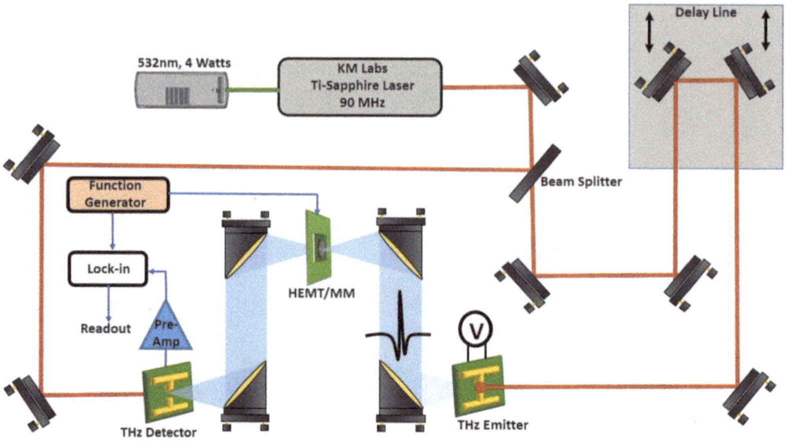

Fig. 4.8 Schematic of the terahertz time-domain spectroscopy (THz-TDS) setup for modulation measurement. Instead of the optical chopper, the THz modulator is used as a frequency selective modulator

$$V_{out}(t) = \alpha \mathbf{E}_0 \cos(\omega_0 t) \tag{4.14}$$

where ω_0 is the resonant frequency of the metamaterial and α is a proportionality constant related to the parameters of the THz-TDS setup, such as the modulation depth of the metamaterial, gain of pre-amplifier and the averaging time-constant of the lock-in amplifier. The assumption that the metamaterial modulates only one frequency is a simplification of the reality, whereas in reality, the metamaterial will modulate a narrow band of frequencies depending on the quality-factor (Q) of the resonance. Therefore, Eq. (4.14) will have the frequencies in that band with their appropriate coefficients.

The time-domain signal $V_{out}(t)$ is plotted in Fig. 4.9a for three different modulation frequencies, 100 kHz, 1 MHz, and 10 MHz. In Fig. 4.9b we plot the spectrum as obtained from Fourier transforming the time-domain data plotted in Fig. 4.9a. It can be seen that the peak of the spectrum lies at 0.46 THz indicating modulation of the metamaterial resonance. Bandwidth of the spectrum remains relatively unchanged between 100 kHz and 10 MHz, and the amplitude of the spectrum increases. At a modulation rate of 10 MHz the bandwidth is observed to decrease a bit and the amplitude falls off from values observed at 1 MHz. The spectrum amplitude has a non-monotonic dependence as a function of frequency that can largely be attributed to the THz-TDS setup itself. THz detection using photoconductive antennas (PCA) has known limitations in THz-TDS at high frequency modulation as documented in past work [35]. Contributions to the degradation of the spectrum amplitude from the HEMT/metamaterial device should remain small as there is several orders of magnitude separation between the gate-to-source modulated voltage and the input noise voltage as reported in HEMT performance studies [36]. In both the

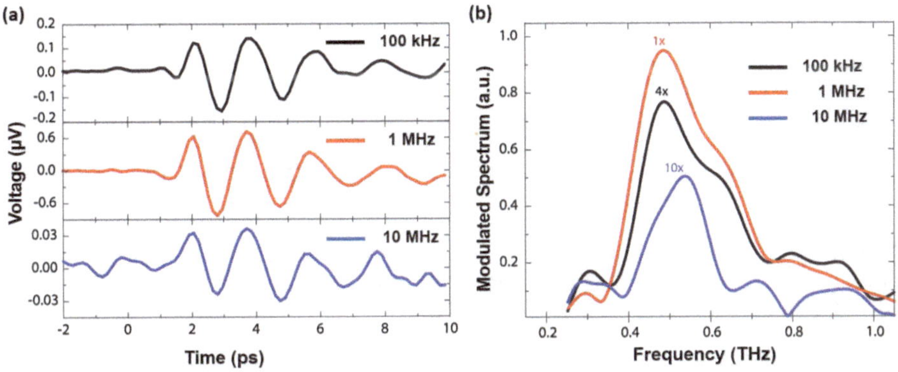

Fig. 4.9 Experimental results from high-speed modulation. (**a**) Time-domain data from the THz-TDS setup for modulation frequencies of 100 Hz, 1 MHz and 10 MHz. (**b**) Spectra calculated by applying Fourier transform to the time-domain data

time-domain signal and the spectrum, the modulation amplitude falls off at higher frequencies and, for this device, the limit is in the neighborhood of 10 MHz. We attribute this as being primarily due to the parasitic capacitance and inductance in the chip assembly and, importantly, not a limitation of either the HEMT or metamaterial device.

References

1. T. Kleine-Ostmann, T. Nagatsuma, J. Infrared Millimeter Terahertz Waves **32**(2), 143 (2011)
2. R. Piesiewicz, T. Kleine-Ostmann, N. Krumbholz, D. Mittleman, M. Koch, J. Schoebel, T. Kurner, IEEE Antennas Propag. Mag. **49**(6), 24 (2007)
3. J. Federici, L. Moeller, J. Appl. Phys. **107**(11), 111101 (2010)
4. G.P. Williams, Rep. Prog. Phys. **69**(2), 301 (2006)
5. H.T. Chen, J.F. O'Hara, A.K. Azad, A.J. Taylor, R.D. Averitt, D.B. Shrekenhamer, W.J. Padilla, Nat. Photonics **2**(5), 295 (2008)
6. H.T. Chen, W.J. Padilla, J.M.O. Zide, A.C. Gossard, A.J. Taylor, R.D. Averitt, Nature **444**(7119), 597 (2006)
7. T. Driscoll, S. Palit, M.M. Qazilbash, M. Brehm, F. Keilmann, B.G. Chae, S.J. Yun, H.T. Kim, S.Y. Cho, N.M. Jokerst, D.R. Smith, D.N. Basov, Appl. Phys. Lett. **93**(2), 024101 (2008)
8. H.T. Chen, S. Palit, T. Tyler, C.M. Bingham, J.M.O. Zide, J.F. OHara, D.R. Smith, A.C. Gossard, R.D. Averitt, W.J. Padilla, N.M. Jokerst, A.J. Taylor, Appl. Phys. Lett. **93**(9), 091117 (2008)
9. O. Paul, C. Imhof, B. Lgel, S. Wolff, J. Heinrich, S. Hfling, A. Forchel, R. Zengerle, R. Beigang, M. Rahm, Opt. Express **17**(2), 819 (2009)
10. H.T. Chen, W.J. Padilla, M.J. Cich, A.K. Azad, R.D. Averitt, A.J. Taylor, Nat. Photonics **3**(3), 148 (2009)
11. L. Mller, J. Federici, A. Sinyukov, C. Xie, H.C. Lim, R.C. Giles, Opt. Lett. **33**(4), 393 (2008)
12. W.L. Chan, H.T. Chen, A.J. Taylor, I. Brener, M.J. Cich, D.M. Mittleman, Appl. Phys. Lett. **94**(21), 213511 (2009)
13. T. Kleine-Ostmann, P. Dawson, K. Pierz, M. Koch, Appl. Phys. Lett. **84**(18), 3555 (2004)

14. C.Y. Chen, T.R. Tsai, C.L. Pan, R.P. Pan, Appl. Phys. Lett. **83**(22), 4497 (2003)
15. T. Kleine-Ostmann, K. Pierz, G. Hein, P. Dawson, M. Marso, M. Koch, J. Appl. Phys. **105**(9), 093707 (2009)
16. D. Shrekenhamer, S. Rout, A.C. Strikwerda, C. Bingham, R.D. Averitt, S. Sonkusale, W.J. Padilla, Opt. Express **19**(10), 9968 (2011)
17. S. Rout, S.R. Sonkusale, APL Photon. **1**(8), 086102 (2016)
18. D. Schurig, J. Mock, D.R. Smith, Appl. Phys. Lett. **88**(4), 041109 (2006)
19. W.J. Padilla, M.T. Aronsson, C. Highstrete, M. Lee, A.J. Taylor, R.D. Averitt, Phys. Rev. B **75**(4), 041102 (2007)
20. B.J. Arritt, D.R. Smith, T. Khraishi, J. Appl. Phys. **109**(7), 073512 (2011)
21. R. Liu, T.J. Cui, D. Huang, B. Zhao, D.R. Smith, Phys. Rev. E **76**(2), 026606 (2007)
22. T.H. Lee, *The Design of CMOS Radio-Frequency Integrated Circuits* (Cambridge University Press, Cambridge, 2004)
23. J.D. Jackson, *Classical Electrodynamics*, 3rd edn. (Wiley, New York, 1998)
24. P.C. Chao, M. Shur, R. Tiberio, K. Duh, P. Smith, J. Ballingall, P. Ho, A. Jabra, IEEE Trans. Electron Devices **36**(3), 461 (1989)
25. Gregory Snider. http://www3.nd.edu/~gsnider (2015)
26. E. Batke, D. Heitmann, Infrared Phys. **24**(2–3), 189 (1984)
27. P.J. Burke, I.B. Spielman, J.P. Eisenstein, L.N. Pfeiffer, K.W. West, Appl. Phys. Lett. **76**(6), 745 (2000)
28. D. Shrekenhamer, J. Montoya, S. Krishna, W.J. Padilla, Adv. Opt. Mater. **1**(12), 905 (2013)
29. J.F. OHara, J.M.O. Zide, A.C. Gossard, A.J. Taylor, R.D. Averitt, Appl. Phys. Lett. **88**(25), 251119 (2006)
30. P.U. Jepsen, R.H. Jacobsen, S.R. Keiding, JOSA B **13**(11), 2424 (1996)
31. W. Knap, J. Lusakowski, T. Parenty, S. Bollaert, A. Cappy, V.V. Popov, M.S. Shur, Appl. Phys. Lett. **84**(13), 2331 (2004)
32. W. Knap, Y. Deng, S. Rumyantsev, J.Q. L, M.S. Shur, C.A. Saylor, L.C. Brunel, Appl. Phys. Lett. **80**(18), 3433 (2002)
33. M. Dyakonov, M. Shur, Phys. Rev. Lett. **71**(15), 2465 (1993)
34. M. Dyakonov, M. Shur, IEEE Trans. Electron Devices **43**(3), 380 (1996)
35. Y. Cai, I. Brener, J. Lopata, J. Wynn, L. Pfeiffer, J.B. Stark, Q. Wu, X.C. Zhang, J.F. Federici, Appl. Phys. Lett. **73**(4), 444 (1998)
36. R. Plana, L. Escotte, O. Llopis, H. Amine, T. Parra, M. Gayral, J. Graffeuil, IEEE Trans. Electron Devices **40**(5), 852 (1993)

Chapter 5
A Terahertz Spatial Light Modulator for Imaging Application

This chapter covers another case study of an all solid-state metamaterial based terahertz (THz) spatial light modulator (SLM) using the HEMT embedded metamaterial based modulator described previously. The motivation behind such a design is first introduced followed by an introduction to single-pixel imaging. Then, the design and assembly details of the spatial light modulator are discussed. The SLM is demonstrated by raster scanning a 6×6 image of an occluded metal object behind a thick polystyrene screen using a single-pixel THz imaging setup.

5.1 Introduction to Single-Pixel Imaging

In recent years, terahertz (THz) imaging has captured significant interest because of its potential for wide range of applications such as bio-detection [1], security screening [2], illicit drug detection [3], and skin cancer detection [4], to name a few. In spite of the intense research activity, simple and low-cost THz imaging systems are still not a reality. Unlike optical imaging, applying an isomorphic mapping of the object on to a THz focal plane array is still a complex and expensive problem. The primary reason for the difficulty is the lack of suitable materials for construction of imaging devices in the terahertz regime (0.1–10 THz, $\lambda = 3$ mm–30 μm), so-called THz Gap [5]. Although real-time terahertz imaging has been demonstrated using focal plane arrays [6, 7], they tend to have higher complexity and operational cost. A different approach is becoming more practical for THz imagers where a single-pixel detector is used instead and the incident THz wave is spatially modulated at the image plane through a coded aperture based on an advanced signal processing theory called compressed sensing (CS) [8, 9]. Such a system, which is relatively compact and low-cost, is schematically illustrated and explained in Fig. 5.1.

The key to single-pixel imaging is to come up with a smallest set of coded masks at the SLM plane, where the image is reconstructed from those encoded single-pixel measurements [8–10] (for details, see Sect. 5.1). The electronically

S. Rout, S. Sonkusale, *Active Metamaterials*, DOI 10.1007/978-3-319-52219-7_5

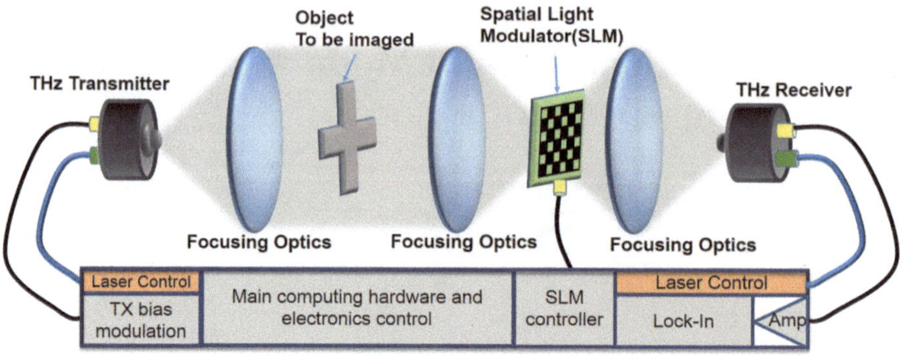

Fig. 5.1 Schematic of a compact and low-cost single-pixel THz imaging system. A difference frequency terahertz (THz) source generates a continuous THz wave which is collimated using focusing optics which illuminates an object that is focused on to a metamaterial based Spatial Light Modulator (SLM). Using compressed sensing (CS) techniques, coded apertures are generated using the electronically controlled SLM. The partially transmitted THz wave is then focused on to a THz receiver antenna which measures the aggregate power of the incident THz wave. Computational imaging approach is employed to reconstruct the image from coded aperture measurements

controlled SLM avoids any mechanical movements as in raster scan thus further improving the speed of acquisition. Although compressed sensing (CS) techniques reduce the number of scans significantly, multiple scans are still required for image reconstruction. Therefore, it is critical for the SLM to have a high switching speed with low-voltage operation that can preferably be integrated within a system-on-chip (SoC) to achieve the speed, power and cost objective for such a system. Significant amount of research effort has been focused in the area of spatial light modulators (SLM) including SLMs based on digital micromirror devices (DMD) [11], micro-electromechanical systems (MEMS) [12], liquid crystal on silicon (LCOS) [13], and graphene-metal plasmonic antennas [14]. These devices perform well for frequencies in the infrared (IR) to visible range and are lossy and incompatible for THz frequencies. These devices can be used in THz SLM applications by creating optical patterns on THz transmissive devices by a reflection geometry [10]. The disadvantage of such a setup is higher complexity and cost.

Recent research has demonstrated THz SLMs using all-electronic dynamic metamaterials used in single-pixel THz imaging systems [10, 15, 16]. These dynamic metamaterial SLMs are electronically controlled by injecting charges or depleting them in the bulk semiconductor substrate, which is an inherently slow process due to the intrinsically large capacitance of what is essentially a large bulk diode structure [15, 17, 18]. This results in a slow switching speed and large dynamic power consumption. Moreover, in typical integrated circuit (IC) processes, this large diode needs 15–50 V to completely deplete the charges. Finally, in such bulk-bias system, on-chip circuits cannot be used making it incompatible with IC integration.

In this chapter, an all solid-state terahertz SLM is demonstrated. The SLM is built from the metamaterial described in Chap. 4, electrically controlled using pseudomorphic high-electron mobility transistor (pHEMT) that is embedded in the split gap of each unit cell [19]. Using a high-speed transistor instead of a bulk semiconductor for active electronic control, we achieve much higher speed of modulation (~10 MHz) and consume much lower power.

5.1.1 A Brief Historical Perspective

The principle behind a single-pixel imager was first demonstrated by Golay in 1949 [21], as a method to achieve faster acquisition rate and higher signal-to-noise ratio (SNR) for infrared (IR) spectrometry. In this remarkable work, he demonstrated the use of modulating mask pattern to measure the spectral information in a dispersive IR spectrometer. His work was way ahead of time when he realized the importance of orthogonal binary digital codes more than two decades before digital computers were to become standard equipment in many laboratories.

Some 20 years after Golay's work, there was revival of his instrumentation techniques but it was not until 1976, until the development of modern computer, that the work by Swift et al. demonstrated the first single-pixel imager using binary optical masks based on Hadamard transform [22], what is known today as compressed sensing [8, 9]. The block diagram of that spectrometer is shown in Fig. 5.2. As shown in the block diagram, this is a spatial spectrometer which captures both spatial and spectral information. Therefore, there two sets of Hadamard masks used in this spectrometer: first set at the entry of the spectrograph to decode the spatial information and the second set at the dispersed plane where the spectral information is dispersed on to a spatial plane. A detailed principle of the imaging theory will be explained next.

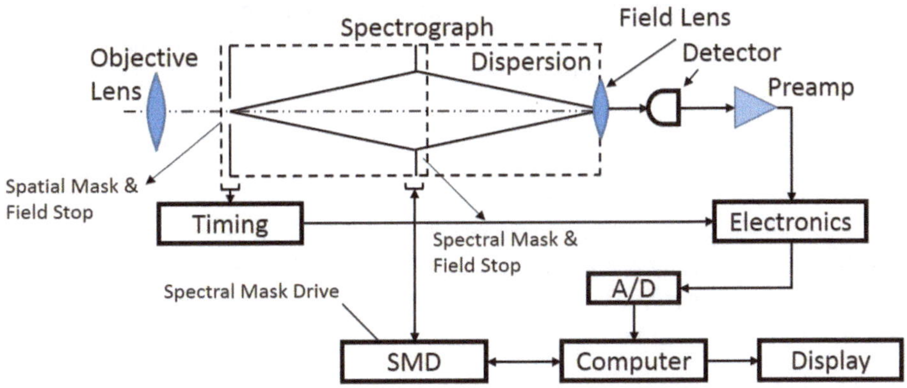

Fig. 5.2 Block diagram of one of the first single-pixel imaging instrument to use binary optical masks based on Hadamard transforms [20]

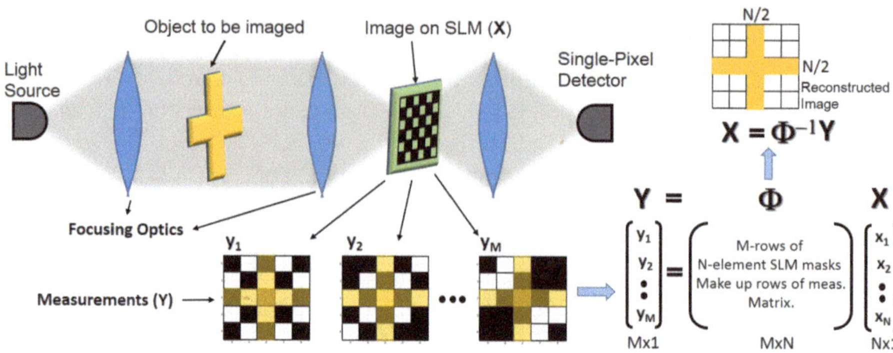

Fig. 5.3 Schematic depiction of the single-pixel multiplex imaging process where the SLM is used to spatial modulate N-pixel image with M-set of known masks and reconstruct the image from the M single-pixel measurements

5.1.2 Imaging Theory

Typical single-pixel imaging setup involves a light source, imaging optics, a single-pixel detector and a spatial light modulator (SLM) as illustrated in Fig. 5.3. Light from an illuminated object is focused on to an SLM which, for binary encoding, selectively passes light on to the single-pixel detector. A sequence (M) of known patterns are displayed on the SLM and single measurements are taken using the detector for each mask or coded aperture resulting in M sets of *single-valued* measurements. Prior knowledge of these masks allows for the reconstruction of a N-pixel image from M measurements. Significant research has been done since the late 1970s in the field of signal processing and information theory to compress the amount of measurements such that $M \ll N$.

The choice of the type of SLM is usually dependent on the type of the single-pixel detector. For an imaging system with intensity based single-pixel detector, binary masks that modulate the intensity are well suited. The field of compressed sensing/imaging is mostly dominated by random and Hadamard based masks [8, 9, 20]. Random binary masks are comprised of 1's and 0's determined by standard random distributions such as Gaussian and Bernoulli. Typically, Hadamard masks are square matrices comprised of $(1, -1)$ where each row is orthogonal to all other rows, but Hadamard matrices comprised of $(-1, 0, 1)$ and $(0, 1)$ are also used and it is proven in [20] that the average mean square error for the best mask of $(0, 1)$ is about four times that for the best mask of $(-1, 0, 1)$. Binary masks have also been created using S-matrix, created by omitting the first row and column of the corresponding normalized Hadamard matrix, substituting $(-1, 1)$ with $(1, 0)$. Using of S-matrix was demonstrated in a terahertz single-pixel imaging experiment in [10].

If we define a one- or two-dimensional image by a vector $\mathbf{X}$ with N-pixels, we do a single measurement y_j for each set of mask with N-elements, the measurements can be expressed in a matrix form

$$\mathbf{Y} = \Phi\mathbf{X} \tag{5.1}$$

where **Y** the column vector with M-elements representing the M measurements and Φ is the $M \times N$ matrix representing the M N-element masks used to make the measurements. For the trivial case of raster scanning the entire image, Φ is an $N \times N$ identity matrix. For a well-conditioned measurement matrix Φ, the reconstructed image **X** can simply be expressed in the matrix form

$$\mathbf{X} = \Phi^{-1}\mathbf{Y} \tag{5.2}$$

Thus, utilizing this new sampling paradigm of compressed sensing (CS), an image is reconstructed, with a bounded error, using fewer measurements than typically needed for a given resolution determined by the Nyquist theorem (N) i.e., $M < N$ [23]. The underlying premise which makes CS possible is that most images of interest are sparse in nature.

Even after decades of research, two major problems which still plaque single-pixel compressive imaging: first, it is computationally very intensive due to its inherent nature of being an NP-hard optimization problem; second, although the number of measurements is greatly reduced, it is still a serial process making video-rate imaging a great technical challenge. Next we look at few research works using spatial light modulators for single-pixel terahertz imaging.

5.2 A Review of THz Spatial Light Modulators

Since the first demonstration of compressive imaging with single-pixel system, there has been significant interest in extending the technology to larger wavelength in the terahertz (THz) regime due to lack THz focal plane arrays. However, due to lack of commercial solid-state spatial light modulators (SLM), early research on THz single-pixel imaging had to resort to mechanically scanned apertures [2, 3, 24]. More recently, metamaterial based SLMs are showing promise in single-pixel imaging [10, 15–17, 25, 26].

A single-pixel terahertz imaging setup using CW pumped high-resitivity silicon (ρ-Si) as a spatial light modulator was demonstrated in [10], a schematic of the setup shown in Fig. 5.4. As shown in the schematic, a collimated LED source is spatially modulated with the desired coded masks using a commercial digital micro-mirror device (DMD) and focused on to the ρ-Si to create the THz SLM. The principle of operation lies in the fact that CW pumped ρ-Si behaves as THz modulator due to free carrier generation and linear recombination in the semiconductor changing the complex dielectric constant that can be described by the Drude model [27]

$$\epsilon(\omega) = \epsilon_{\infty} - \frac{\omega_p^2}{\omega(\omega + i\Gamma)} \tag{5.3}$$

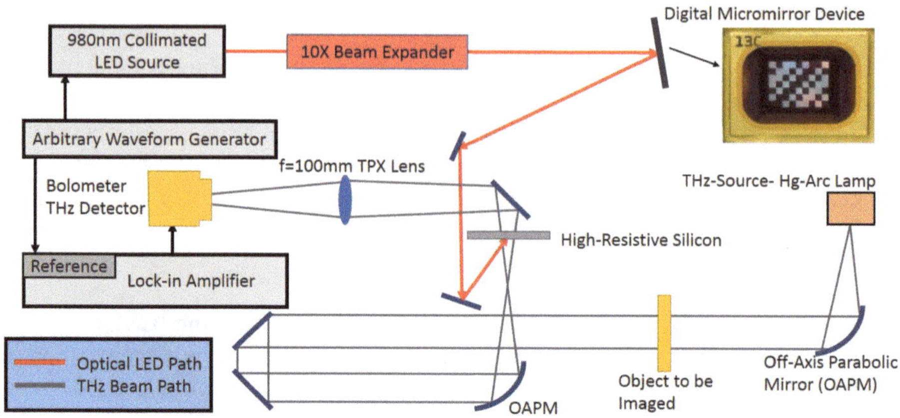

Fig. 5.4 Schematic of a single-pixel terahertz imaging setup using CW pumped high-resitivity silicon as a spatial light modulator [10]

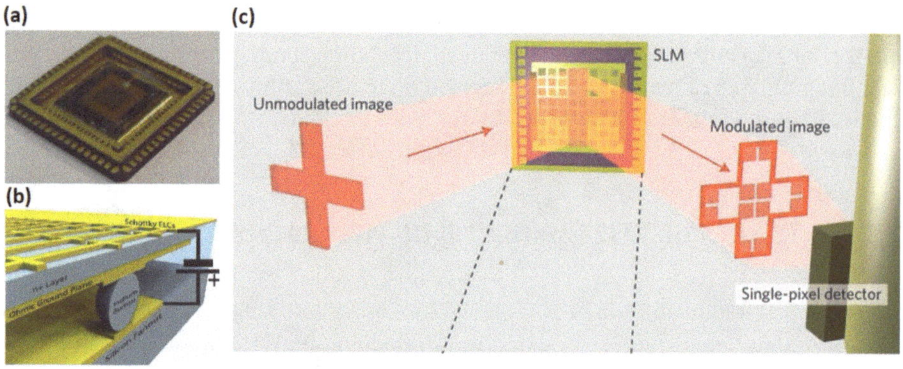

Fig. 5.5 (**a**), (**b**) Design and structure detail of the electronically controlled THz metamaterial absorber based spatial light modulator (MMA-SLM) [25]. (**a**) Image of MMA-SLM as assembled in chip carrier package. (**b**) Cross-sectional schematic view of a single pixel. (**c**) Schematic of a THz compressive imaging setup using the SLM in (**a**), (**b**) [16]

A fundamental limitation of this technique is the carrier lifetime of ρ-Si of $\tau = 25\,\mu\text{s}$ limits the switching speed to about 10 kHz. Moreover, the LED source, DMD, and the optics to create the spatial CW pumping make the system complex and expensive.

Figure 5.5a, b shows the SLM based on multi-resonant, electronically controlled metamaterial absorber [25]. The overall SLM system architecture is shown schematically in Fig. 5.5b and consists of metamaterial absorber pixels flip chip bonded to a Silicon chip carrier with routing to bond pads which are wire-bonded to a leadless chip carrier (LCC). The metamaterial absorber consists of two metallic layers with a dielectric spacer lying in-between. The top metal layer is patterned in order to respond resonantly to the electric component of an incident electromagnetic wave. A bottom ground plane layer is spaced relatively close to the top layer, thus allowing the external magnetic field to couple, as shown in Fig. 5.5b. The SLM was demonstrated in a single-pixel compressive imaging experiment using a reflection geometry [16], a schematic of the setup shown in Fig. 5.5c. The same SLM has also been used in recent works involving single-pixel quadrature imaging [28] and frequency-division-multiplexed single-pixel imaging [29]. This electrically controlled SLM is a great improvement over the optically pumped SLM allowing for simpler and cheaper THz imagers. However, having to use a certain height of dielectric to design for a desired resonant frequency for the absorber makes the design very rigid and precludes the use of commercial foundry process where the vertical dimensions are fixed. Additionally, reflection geometry imaging setups add to the complexity and hence the cost.

One of the first electronically controlled, all solid-state terahertz SLM was demonstrated by Chan et al. [15] in 2009 using an active metamaterial design demonstrated earlier by Chen et al. [17] in 2006. As shown in Fig. 5.6b, c, the modulator is based on the principle of shunting the metamaterial split gap by injecting carrier in the substrate using an external bias voltage between the metamaterial and the substrate. Figure 5.6a shows the SLM assembly of 4×4 array of individual pixels. Each pixel is independently controlled by an external voltage between the $1 \times 1\,\text{mm}^2$ Schottky electric pad and the ohmic contact. The SLM was characterized in a transmissive geometry at 0.36 THz with modulating voltage of 16 V at switching speed of kilo-hertz. Although this work marked the beginning

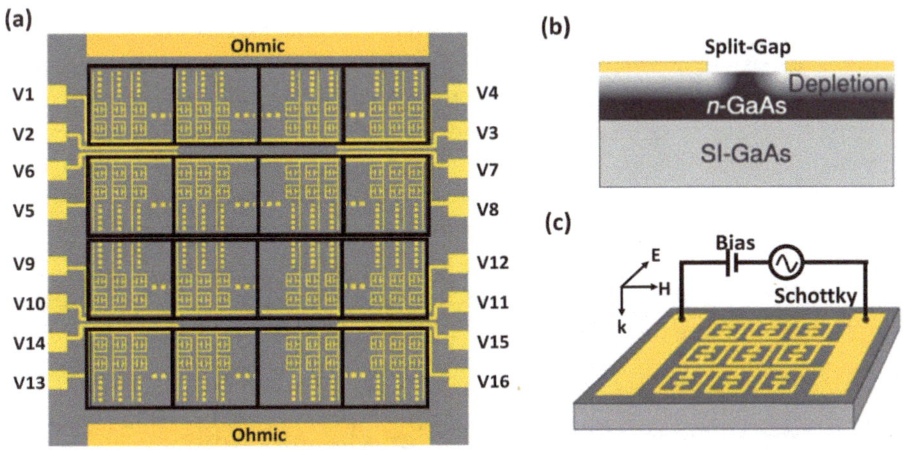

Fig. 5.6 Schematic of a solid-state spatial light modulator for terahertz frequencies (0.36 THz). (**a**) The terahertz SLM (not drawn to scale), a 4×4 array of individual pixels shown in (**c**). (**b**) Substrate and the depletion region near the split gap of a single SRR, where the gray scale indicates the free charge carrier density. (**c**) Each single pixel on the metamaterial based terahertz SLM. A voltage bias (16 V) between the $1 \times 1\,\text{mm}^2$ Schottky electric pad and the ohmic contact controls the substrate charge carrier density near the split gaps, tuning the strength of the resonance [15]

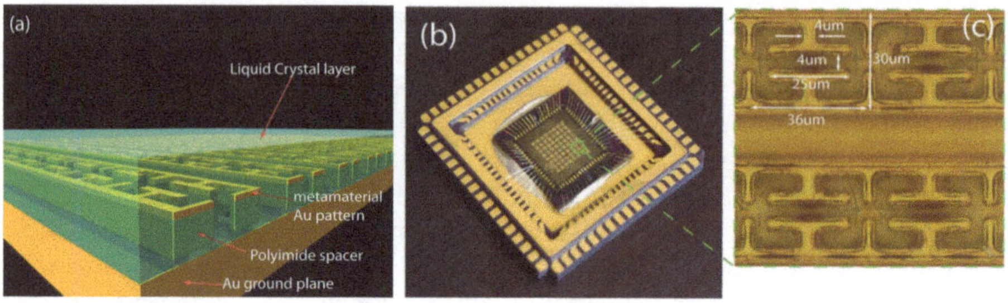

Fig. 5.7 Liquid Crystal (LC) Metamaterial Absorber (MMA) Spatial Light Modulator (SLM) for THz Applications. (**a**) 3D cross-section schematic of the MMA array covered with LC. (**b**) Picture of the MMA SLM device coated with LC. (**c**) Close-up of MMA unit cells [30]

of solid-state THz SLMs, there are few drawbacks associated with the bias voltage being applied across the entire substrate. Biasing the entire substrate precludes it from being used on systems-on-chip (SoC) applications, a main attraction of solid-state SLMs. The switching speeds are slow due to the large associated capacitance of the substrate and the switching voltage is high related to the breakdown voltage of the substrate.

Although advances have been made in liquid crystal on silicon (LCOS) spatial light modulator technology [13], there applications in terahertz have been limited. More recently, metamaterial absorbers embedded in liquid crystals were demonstrated in reflection geometry as terahertz spatial light modulators [30], Fig. 5.7. As shown in Fig. 5.7a, the liquid crystal (LC) forms the dielectric in the split gap of the resonator. By applying a bias voltage and thus electric field across the LC, the polarization of the LC is changed which in turn changes the resonant frequency of the absorber, resulting in voltage controlled modulation. This work showed the viability of using a liquid crystal with metamaterial absorbers with results of 75% modulation depth at 3.76 THz. Although promising, the use of 15 V switching voltage, modulation speeds of only 1 kHz and ~70% signal absorbtion are significant drawbacks of the technology to make it a serious contender for THz SLM.

In the next sections we detail the design and characterization of a terahertz SLM that aims to solve some the major problems stated in the previous works.

5.3 Spatial Light Modulator Design and Assembly

The basic principle of constructing a terahertz SLM is an array of terahertz devices that can independently modulate the transmission (or reflection) of terahertz wave at their respective array positions, thus spatially modulating the transmissive (or reflective) wave front.

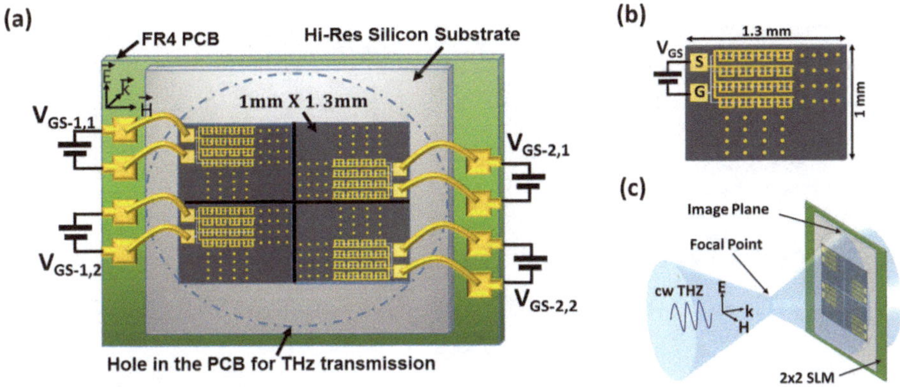

Fig. 5.8 SLM design and assembly details. (**a**) The terahertz SLM is a 2 × 2 array of individual pixels in (**b**). Each pixel is independently controlled by an external voltage source $V_{GS-i,j}$ (0–1 V) to control the transmission of the incident THz wave. (**b**) Each single pixel on the terahertz SLM contains a 1 × 1.3 mm^2, 551 element array of metamaterial ELCs. These elements are connected together with metal wires and the gates are all connected in a separate metal layer to serve as the source and gate connection of the pHEMTs, respectively. An external voltage bias V_{GS} controls the charge carrier density near the split gaps, tuning the strength of the resonance. (**c**) Experimental configuration for THz transmission imaging through the SLM. The SLM is placed at the image plane to spatially modulate the transmissive wave front

The terahertz modulator described in Sect. 4.2 is used to construct the SLM. This metamaterial consists of a planar array of sub-wavelength-sized electric-LC (ELC) resonators fabricated using a matured commercial GaAs technology [see Figs. 4.1a and 5.8b], operating at room temperature using a low supply voltage (1 V), and consuming less than 1 mW of power.

The principle of modulation is based on electrical control of the metamaterial resonance (0.46 THz), which in turn changes the damping factor of the Drude-Lorentz form of effective permittivity for the metamaterial. The control of the metamaterial resonance is realized by changing the channel carrier density of pseudomorphic high-electron mobility transistors (pHEMTs) that are embedded in each of the split gaps of the metamaterial elements, which in turn changes the loss at the capacitive split gaps and therefore the oscillator strength of all of the individual ELC elements within a pixel. Because of the localized pHEMT in the split gap, the device has been demonstrated for fast modulation [19] (∼10 MHz) that is capable of much higher speed with proper design. Compared to other electrically controlled metamaterial [17, 31], where the whole substrate is used to control the resonance, this device offers element-level control that offers higher switching speed and offers future opportunity of creating more exotic devices. A metamaterial based SLM also offers the advantage of creating hyper-spectral imaging by creating arrays with different frequencies [25]. A detailed description of the metamaterial design can be found in Sect. 4.2. Figure 5.9 shows close-up pictures of the metamaterial device at various magnification ratio.

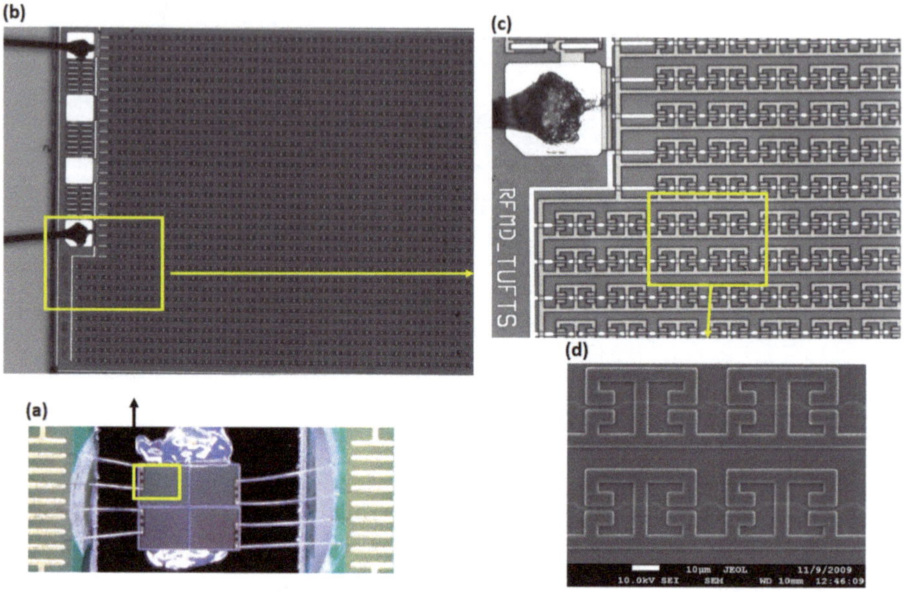

Fig. 5.9 (**a**) Close-up picture of the SLM assembly on a Hi-res silicon that is mounted on PCB and wire bonded to copper traces on the PCB. (**b**) Magnified picture of one the dies in the region shown in the yellow box in (**a**) below (**c**) Magnified picture of the die near a pad showing all the drain/source connection and all the gates connected to the pad. (**d**) A SEM picture of the metamaterial structure

To demonstrate the terahertz SLM, a 2 × 2 pixel device was constructed, where each pixel is $1.0 \times 1.3\ \text{mm}^2$ with 551 elements with an active area of $0.88 \times 1.12\ \text{mm}^2$ for each pixel as shown in Fig. 5.8b. The 2 × 2 array is die-attached (only at the corners) to a high-resistive silicon (ρ-Si) substrate that is mounted on an FR-4 based PCB with a hole underneath the ρ-Si to allow THz transmission as shown in Fig. 5.8a. Figures 5.9a and 5.12 show the SLM and printed circuit board (PCB) assembly. All the gates of the HEMTs for each pixel are connected to a 100 µm × 100 µm bond pad and all the metamaterials for each pixel are connected to a separate bond pad which provides the DC bias for the drain/source connection. These bond pads are bonded out to the test circuit board such that each pixel can be biased (V_{GS}) independently thus enabling a 2 × 2 electrically controlled terahertz SLM. Figures 5.10 and 5.11 show the architecture and circuitry for electronic control of the SLM. Figure 5.8c illustrates the experimental configuration of the SLM with respect to the focal point and the image plane of the imaging optics of the setup shown in Fig. 5.13. The SLM is placed at the image plane to be able to spatially modulate the wave front.

Please note that the 2×2 size of the SLM is purely a limitation for reasons of cost and assembly logistics. The cost of fabrication limited our die size to $1.0 \times 1.3\ \text{mm}^2$ and the SLM assembly of 2 × 2 array enabled a simpler assembly process. Using this technology, an independently controlled array of an arbitrary size can be easily implemented, only limited by the yield of the fabrication technology.

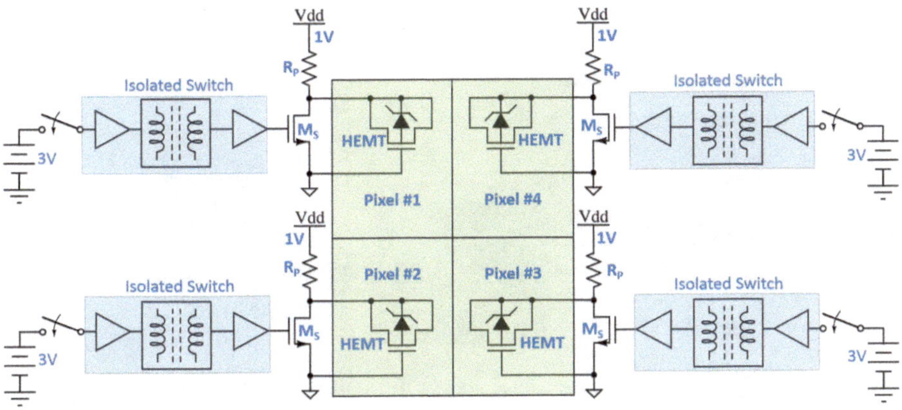

Fig. 5.10 A *top level view* of the circuit architecture to control the SLM

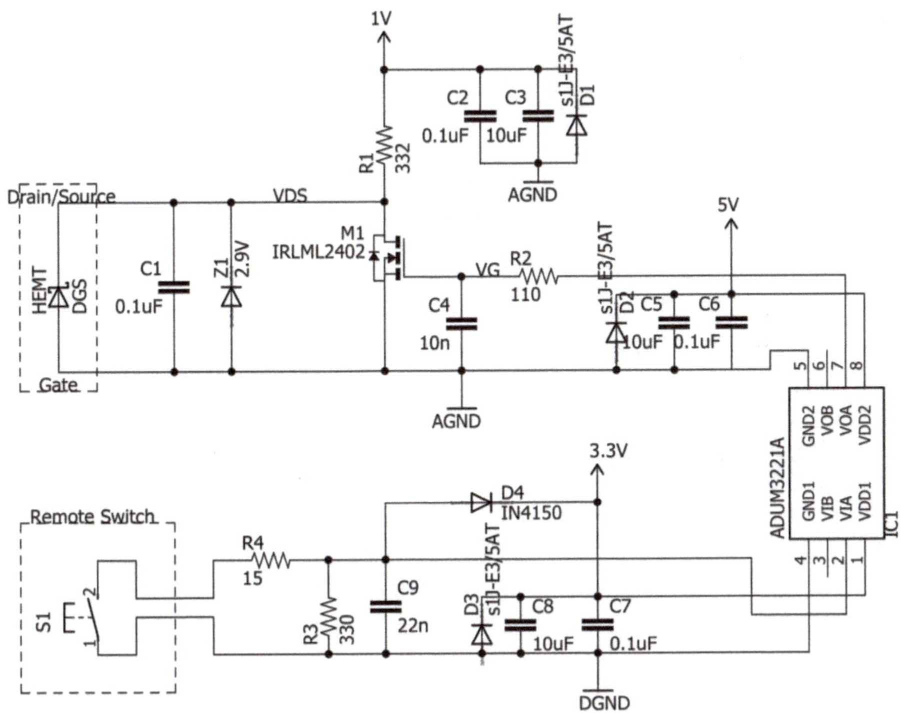

Fig. 5.11 Details of the circuit, including the part numbers and the values, used to control one of the SLM's pixels

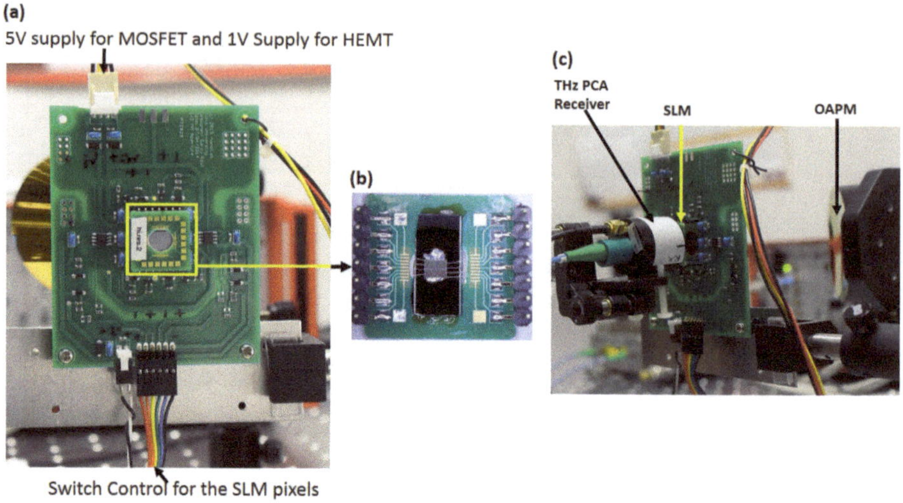

Fig. 5.12 (**a**) The main printed circuited board (PCB) for electronic control of the SLM that is mounted using a detachable PCB in the middle. (**b**) The underside of the detachable PCB is shown with the 2 × 2 SLM tile mounted on a Hi-Res silicon. (**c**) The SLM position in the experimental setup is shown with respect to the THz receiver and the off-axis parabolic mirror (OAPM)

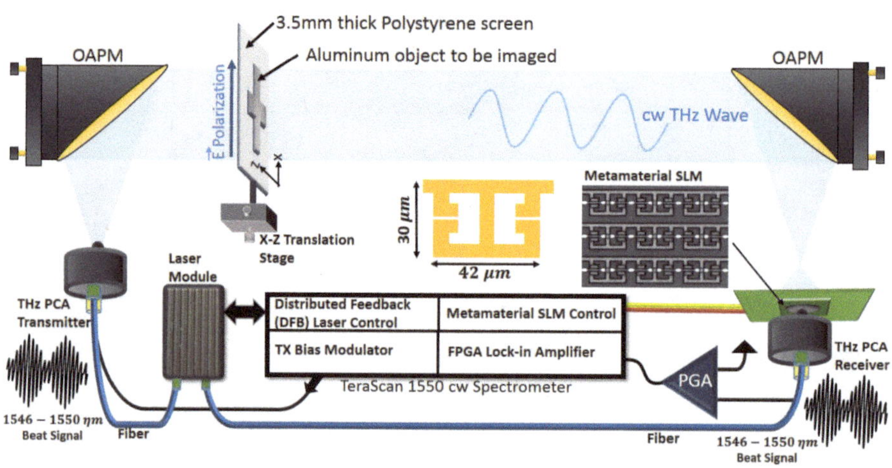

Fig. 5.13 Experimental setup for terahertz characterization and imaging. Schematic diagram of the continuous-wave (cw) terahertz setup for characterizing the metamaterial SLM. The magnified picture of the metamaterial is shown in the inset and the yellow overlay shows the geometry of each unit cell. A fiber-coupled photo-conductive antenna generates the THz wave from laser beat signal that is collimated and focused by a pair of Off-Axis Parabolic Mirrors (OAPMs). The object, an aluminum metal cross on a 3.5 mm thick polystyrene screen is placed in the path of the collimated beam. The metamaterial SLM is placed 15 mm away from the focal point and the single-pixel THz detector is placed right behind the SLM. The receiver photocurrent is first amplified by a programmable gain amplifier (PGA) and then lock-in detected by a custom FPGA [32]

5.4 Circuit Design for Electronic Control of the SLM

Figure 5.10 shows the top level architecture of the circuit to electronically control the SLM. The VGS of all the embedded HEMTs in a pixel is controlled by an identical circuit. The VGS of all the HEMTs are represented by one device in the figure essentially all schottky diodes in parallel. The VGS is switched between 0 V and −1 V using the MOSFET M_S and the pull-up resistor R_P. In order to switch "on" a pixel ($V_{GS} = -1$ V), the MOSFET M_S for that particular pixel is switched off which pulls drain/source of the HEMTs to Vdd (1 V) and to switch "off" the pixel ($V_{GS} = 0$ V), the MOSFET is turned on and the drain/source of the HEMTs are pulled down to ground. The gate of the MOSFETS are driven by an isolated switch to eliminate any ground noise from the switching side since that is controlled by the computer through long wires creating lot of ground bounce. The details of the control circuit are shown in the next figure.

Figure 5.11 shows the details of the circuit, including the part numbers and the values, used to control one of the SLM's pixels. All the HEMTs of a pixel are represented by the schottky diode DGS. The zener diode Z1 and capacitor C1 are used to filter any large transients during the switching to avoid damaging the HEMTs. M1 and R1 are used to switch the VGS of the HEMTs as explained in the previous figure. Capacitor C1 and C2 are used to filter any noise on the 1 V supply. D1 is for protection against wrong polarity. R2 and C4 are used to filter large transients at the gate of the MOSFET. C5, C6, and D2 are used to filter and protect the 5 V supply on the board. IC1 (Analog Devices ADUM3221A) is transformer based isolated switch that blocks any noise in the switch (S1) side since typically that is a noisy domain. Switch S1 is connected to the PCB board via a long wire since this is remote to where the metamaterial SLM is placed. Therefore, R3, R4, C9, and D4 are used to eliminate the large transients and smooth the transitions. C7, C8, and D4 are used to filter the noise and protect the 3 V supply used to power the switching side of the isolator.

Figure 5.12 shows the PCB design for the electronic control of the SLM. Figure 5.12a shows the mounted PCB with all electronics (except for the switches) for controlling the SLM pixels. In the middle of the PCB is the daughter card with the SLM pixels which is mounted on a high-resistivity (Hi-res) silicon substrate with a hole in the PCB for THz transmission tests. The power supplies are well grounded and shielded for noise immunity. Figure 5.12b shows the back side of the daughter card showing the SLM assembly on a Hi-res silicon. The four metamaterial dies are attached using a high-viscosity die attach to avoid any material in between the metamaterial and the Hi-Res silicon. The drain/source and the gate connections for the bond pads are wire bonded to the PCB which is electrically connected to the right-angle pins that is used to insert the daughter card in the main PCB. Figure 5.12c shows the picture of the setup near the THz receiver. It can be seen from the picture that the receiver is placed right behind the SLM as single-pixel detector.

5.5 Experimental Setup for Terahertz Characterization and Imaging

A schematic of the THz imaging and characterization system is shown in Fig. 5.13. The terahertz SLM is characterized in transmission mode using a commercial continuous-wave (cw) THz spectroscopy system, TeraScan 1550 by Toptica Photonics (http://www.toptica.com, 2015). The cw terahertz spectrometer generates linearly polarized THz frequency from 60 GHz to 1.2 THz using a pair of tuned lasers (1546 and 1550 nm). The temperature controlled beat frequency is fiber-coupled to a InGaAs photo-diode with a bow-tie antenna which is bias modulated (± 1.2 V) at 7.629 kHz.

The THz imaging optics consist of two 76.2 mm diameter 90° off-axis parabolic mirrors (OAPMs) each with an effective focal length of $f_L = 152.2$ mm. The source-side OAPM is f_L from the source and back-illuminates the object a distance 200 mm away. The second OAPM, a distance 800 mm from the object, focuses on the radiation and forms a conjugate image on the SLM 167.2 mm away. The system has an overall magnification of 0.24, mapping a 35 mm diameter object to 8.4 mm. Please note, the correct image plane is 184 mm away from the second OAPM with an overall magnification ratio of approximately 0.5. Given limited Signal-to-Noise Ratio (SNR), acquiring image with adequate fidelity at the image plane was challenging.

The choice of our SLM location was based on the optimum SNR vs. resolution. The THz detector (fiber-coupled InGaAs photo-mixer) is placed right behind the SLM to measure the aggregate THz radiation power passing through the SLM. Since the diameter of receiver's semi-hemispherical silicon lens is approximately the size of the SLM, we avoid the second set of optics to give us better fidelity i.e. SNR. The SLM orientated such that the linearly polarized electric field of the THz wave is across the split gaps of the metamaterial elements. The detected photocurrent is pre-amplified using a programmable gain amplifier (PGA) and then lock-in detected ($\tau = 620$ ms) using Toptica's proprietary FPGA module. The detected photocurrent I_{ph} can be expressed as $I_{ph} \propto E_{THz} \cos(\Delta\phi)$ [33] where E_{THz} is the amplitude of the terahertz electric field and $\Delta\phi$ is the phase difference between the terahertz wave and the laser beat signal at the detector. Therefore, the detected photocurrent I_{ph} oscillates with the THz frequency with the period set by the length of the terahertz beam. The frequency response of a sample is calculated by measuring the envelope of the oscillating I_{ph}. The oscillating period limits the frequency resolution, which was $\approx$0.2 GHz for our setup. The basic principle and working of cw-THz spectroscopy is covered in Sect. 3.3.2.

For proof of concept, a 15 mm wide aluminum cross placed on a 3.5 mm thick polystyrene screen was imaged. The polystyrene screen is opaque to visible light but transparent to 0.45 THz. The object was raster scanned 3×3 with a step size of 15 mm and for each raster position the SLM scans a 2×2 image resulting in 6×6 image size with 7.5 mm physical resolution.

5.6 Results and Discussions

5.6.1 Terahertz Characterization of the Spatial Light Modulator

The spatial light modulator (SLM) was characterized and single pixel THz imaging was performed in a transmission geometry using the experimental apparatus described in the previous section (Sect. 5.5, Fig. 5.13). All the characterization was done without the object, with the SLM and the detector placed at the focal point of the OAPM for maximum signal-to-noise ratio (SNR).

Figure 5.14a shows the transmission spectra using the envelope of the detected photocurrent as each of the four pixels is turned "on"($V_{GS-i,j} = -1\,V$) and

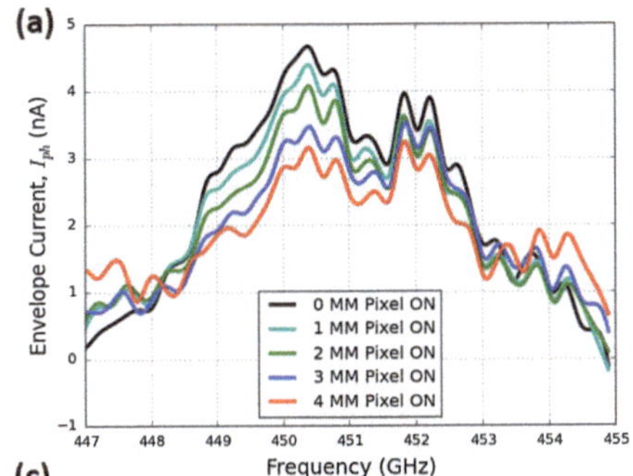

(b)

No. of pixel "ON"	$V_{GS-1,1}$	$V_{GS-1,2}$	$V_{GS-2,2}$	$V_{GS-2,1}$	I_{ph} (450.2 GHz)
0	0 V	0 V	0 V	0 V	4.48 nA
1	-1 V	0 V	0 V	0 V	4.11 nA
2	-1 V	-1 V	0 V	0 V	3.73 nA
3	-1 V	-1 V	-1 V	0 V	3.23 nA
4	-1 V	-1 V	-1 V	-1 V	2.86 nA

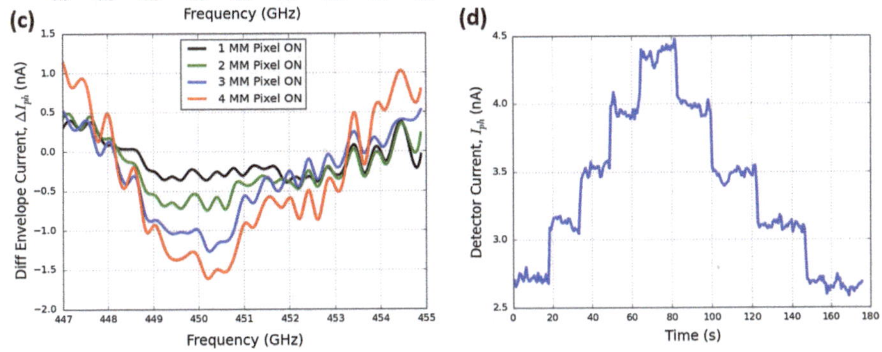

Fig. 5.14 Characterization of the THz spatial light modulator (SLM). (**a**) Transmission spectra (447–455 GHz) of the SLM by measuring the envelope of the detected photocurrent I_{ph} as each pixel is switched "on" sequentially. (**b**) Table of gate-to-source voltage V_{GS} for each pixel of the SLM to have 0, 1, 2, 3, or 4 active ($V_{GS} = -1\,V$) pixels used in the SLM characterization in (**a**), (**c**) and also in the imaging experiment (result shown in Fig. 5.15). The last column tabulates the detector photocurrent at 450.2 GHz extracted from the spectral data in (**a**) demonstrating a total amplitude modulation of 36% (1.62 nA) with an average of 9% (0.41 nA) modulation depth for each pixel. (**c**) The differential transmission spectra (ΔI_{ph}) with reference spectra of maximum transmission, i.e. all the metamaterials are "off" ($I_{ph}(V_{GS} = 0\,V$ for all MM)). (**d**) Real-time detected photocurrent at 450.2 GHz [32]

"off"($V_{GS-i,j} = 0$ V) sequentially in a clockwise or anti-clockwise direction where i and j correspond to the row and column, respectively, as shown in Fig. 5.8a.

Figure 5.14b tabulates the gate-to-source voltage (V_{GS}) for each pixel of the SLM, used to sequentially turn "on" or "off" each pixel for the SLM characterization and imaging as well. The spectra for each SLM configuration is result of a frequency scan from 447 GHz to 455 GHz with a step size of 0.005 GHz and lock-in time-constant of 620 ms. The last column in Fig. 5.14b tabulates the extracted photocurrent from the transmission spectra at $f = 450.2$ GHz. Between all 4 pixels "on" and "off," an amplitude modulation depth of 36% (1.62 nA) is measured and an average of 9% (0.41 nA) modulation depth for each pixel at the designed resonant frequency of 0.45 THz. In order to emphasize the spatial modulation near the resonant frequency, the modulation spectra for each pixel is plotted relative to a reference.

Figure 5.14c shows the differential transmission spectra extracted from the data in Fig. 5.14a. The reference for the differential spectra is when the SLM transmission is maximum, i.e. all the metamaterials are "off" ($I_{ph\_ref}(f) = I_{ph}(f)[V_{GS} = 0$ V for all MM]). Then for each SLM configuration the differential spectra is expressed as $\Delta I_{phN}(f) = I_{phN}(f) - I_{ph\_ref}(f)$ where N denotes one of the four pixels.

Figure 5.14d shows the real-time SLM response to a continuous THz wave of 0.45 THz, further demonstrating the SLM action.

5.6.2 *Single-Pixel Terahertz Imaging*

Single pixel imaging experiments were performed using the experimental setup described in the previous section (Sect. 5.5, Fig. 5.13). For proof of concept, an aluminum cross placed behind a 3.5 mm thick polystyrene screen [See inset in Fig. 5.15a] was imaged. Polystyrene is opaque to visible light but transparent to 0.45 THz, thus an ideal setup for demonstrating THz imaging. For the imaging setup, the spatial resolution that can be resolved at the object plane by a single SLM pixel is approximately 7.5 mm × 7.5 mm. An aluminum cross with an arm width of 15 mm and a length of 45 mm was used, as illustrated in Fig. 5.15c, so the whole object can be resolved with high certainty using a 6 × 6 image size. Since the SLM array is only 2 × 2, the object was raster scanned 3 steps across and 3 steps down with a 15 mm step size. For each step, the SLM was used to scan 2 × 2 pixels for that position thus resulting in a 6 × 6 image with a resolution of 7.5 mm × 7.5 mm. It should be emphasized again that the 2 × 2 size of the SLM is due to reasons of cost and assembly logistics. Raster scanning would not have been necessary if we had implemented a 6 × 6 array of SLM.

To achieve maximum signal-to-noise ratio (SNR), each pixel involves four measurements; two for reference differential photocurrent, $\Delta I_{phN\_ref} = I_{phN\_ref}(V_{GS} = 0\text{ V}) - I_{phN\_ref}(V_{GS} = -1\text{ V})$, where the photocurrents are measured with only the polystyrene screen without the metal cross and, two for the differential photocurrent with the metal cross behind the polystyrene screen, $\Delta I_{phN} = I_{phN\_OFF}(V_{GS} =$

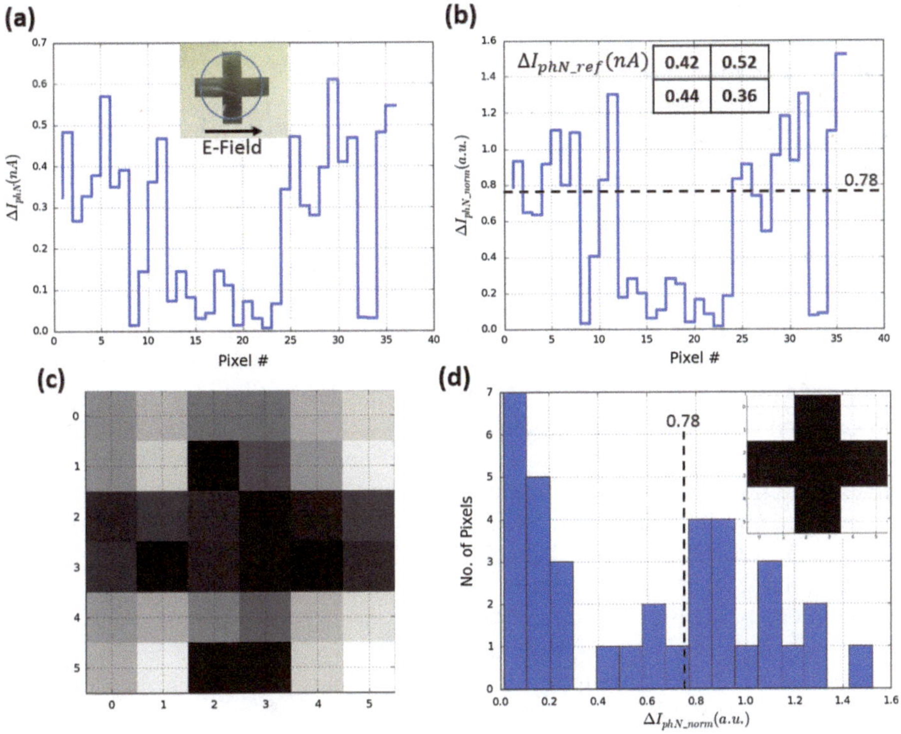

Fig. 5.15 Single-pixel terahertz imaging. (**a**) Differential lock-in current (ΔI_{phN}) as each pixel is turned "on" and "off" for all 36 pixels. Inset of the scanned object: aluminum metal cross of width 15 mm on a 3.5 mm thick polystyrene screen. The linearly polarized field of the THz transmitter is shown with respect to the object. (**b**) Normalized differential lock-in current ($\Delta I_{\text{phN\_norm}}$) where each pixel is normalized to its reference differential lock-in current $\Delta I_{\text{phN\_ref}}$. Inset shows $\Delta I_{\text{phN\_ref}}$ for each of the four pixels. (**c**) $\Delta I_{\text{phN\_norm}}$ is displayed as 6×6 *gray scale* image with a physical resolution of 7.5 mm. (**d**) Histogram of $\Delta I_{\text{phN\_norm}}$. Twenty pixels (object) with $\Delta I_{\text{phN\_norm}} < 0.78$ shown with the *vertical dashed line*. Inset: A binary image is created using the histogram data with a threshold value of $\Delta I_{\text{phN\_norm}} = 0.78$ [32]

0 V) $- I_{\text{phN\_ON}}(V_{\text{GS}} = -1\,\text{V})$, for the *N*th pixel. Then, the measured differential photocurrent (ΔI_{phN}) is normalized with respect to the reference differential photocurrent ($\Delta I_{\text{phN\_ref}}$), which can be expressed as $\Delta I_{\text{phN\_norm}} = \Delta I_{\text{phN}}/\Delta I_{\text{phN\_ref}}$. For each measurement, the frequency was scanned from 450.025 GHz to 450.25 GHz and the peak-to-peak value of the lock-in photocurrent was recorded. This was done to remove any phase information in the measurement. In order to save time, $\Delta I_{\text{phN\_ref}}$ was measured only once for the 4 SLM pixels assuming it doesn't change significantly during the raster scan.

Figure 5.15a displays the differential photocurrent (ΔI_{phN}) for each of the 36 scanned pixels. The normalized differential photocurrent ($\Delta I_{\text{phN\_norm}}$) is plotted in Fig. 5.15b. The reference differential photocurrent ($\Delta I_{\text{phN\_ref}}$) for each of the 4

SLM pixel is tabulated in the inset of Fig. 5.15b. Note, the maximum value of the normalized differential current is greater than one due to drifts in measurements from one scan to scan to another, a source of noise as well.

In order to get a spatial view of the normalized differential photocurrent $\Delta I_{\text{phN\_norm}}$, it is displayed as a gray scale image in Fig. 5.15c. The row and column index of Fig. 5.15c is related to the pixel number in Fig. 5.15a and b by the expression $6 \times i + j + 1$, where i and j are the row and column index of Fig. 5.15c, respectively. The low fidelity in the image quality is primarily due to the noise in the system. As mentioned in the previous section (Sect. 5.5), the SLM and the detector are placed closer to the focal point for measurable signal-to-noise ratio (SNR) at the expense of lower resolution further lowering the fidelity of the image. One way to remove the noise in bi-modal distributed pixels is to create a binary image based on a threshold that is mid way between the two modes of distribution. The histogram of $\Delta I_{\text{phN\_norm}}$ is shown in Fig. 5.15c. Since the two distributions (for the object and the polystyrene screen) are very close to each other due to higher noise in the system, the threshold was chosen based on the a priori knowledge of the object.

Since the object (aluminum cross) should be composed of 20 pixels, the threshold was chosen at the 20th "darkest" pixel, $\Delta I_{\text{phN\_norm}} = 0.78$. Based on this threshold value, a binary image was created as shown in inset of Fig. 5.15d. From the binary image, it can be seen that the aluminum cross placed behind the thick polystyrene screen, which is opaque for visible light, is identified using the single pixel THz imaging setup, demonstrating the viability for security screening, one of many potential applications of THz imaging.

References

1. T.M. Korter, D.F. Plusquellic, Chem. Phys. Lett. **385**(1–2), 45 (2004)
2. N. Karpowicz, H. Zhong, C. Zhang, K.I. Lin, J.S. Hwang, J. Xu, X.C. Zhang, Appl. Phys. Lett. **86**(5), 054105 (2005)
3. K. Kawase, Y. Ogawa, Y. Watanabe, H. Inoue, Opt. Express **11**(20), 2549 (2003)
4. R.M. Woodward, V.P. Wallace, R.J. Pye, B.E. Cole, D.D. Arnone, E.H. Linfield, M. Pepper, J. Invest. Dermatol. **120**(1), 72 (2003)
5. G.P. Williams, Rep. Prog. Phys. **69**(2), 301 (2006)
6. A.W.M. Lee, Q. Qin, S. Kumar, B.S. Williams, Q. Hu, J.L. Reno, Appl. Phys. Lett. **89**(14), 141125 (2006)
7. Z. Jiang, X.C. Zhang, IEEE Trans. Microwave Theory Tech. **47**(12), 2644 (1999)
8. D. Donoho, IEEE Trans. Inf. Theory **52**(4), 1289 (2006)
9. E. Candes, J. Romberg, T. Tao, IEEE Trans. Inf. Theory **52**(2), 489 (2006)
10. D. Shrekenhamer, C.M. Watts, W.J. Padilla, Opt. Express **21**(10), 12507 (2013)
11. D. Dudley, W.M. Duncan, J. Slaughter, Emerging digital micromirror device (DMD) applications, in *Proceedings of SPIE*, vol. 4985, pp. 14–25. doi:10.1117/12.480761; MOEMS Display and Imaging Systems, 14 (January 20, 2003)
12. J. Dunayevsky, D. Marom, J. Microelectromech. Syst. **22**(5), 1213 (2013)
13. W.P. Bleha, L.A. Lei, Advances in Liquid Crystal on Silicon (LCOS) spatial light modulator technology, in *Proceedings of SPIE*, vol. 8736 (2013), pp. 87, 360A–87, 360A–8. doi:10.1117/12.2015973; Display Technologies and Applications for Defense, Security, and Avionics VII, 87360A (June 4, 2013)

14. Z. Li, N. Yu, Appl. Phys. Lett. **102**(13), 131108 (2013)
15. W.L. Chan, H.T. Chen, A.J. Taylor, I. Brener, M.J. Cich, D.M. Mittleman, Appl. Phys. Lett. **94**(21), 213511 (2009)
16. C.M. Watts, D. Shrekenhamer, J. Montoya, G. Lipworth, J. Hunt, T. Sleasman, S. Krishna, D.R. Smith, W.J. Padilla, Nat. Photonics **8**, 605 (2014)
17. H.T. Chen, W.J. Padilla, J.M.O. Zide, A.C. Gossard, A.J. Taylor, R.D. Averitt, Nature **444**(7119), 597 (2006)
18. B. Sensale-Rodriguez, R. Yan, M.M. Kelly, T. Fang, K. Tahy, W.S. Hwang, D. Jena, L. Liu, H.G. Xing, Nat. Commun. **3**, 780 (2012)
19. D. Shrekenhamer, S. Rout, A.C. Strikwerda, C. Bingham, R.D. Averitt, S. Sonkusale, W.J. Padilla, Opt. Express **19**(10), 9968 (2011)
20. R.D. Swift, R.B. Wattson, J.A. Decker, R. Paganetti, M. Harwit, Appl. Opt. **15**(6), 1595 (1976)
21. M.J.E. Golay, JOSA **39**(6), 437 (1949)
22. N.J.A. Sloane, M. Harwit, Appl. Opt. **15**(1), 107 (1976)
23. E.J. Candes, M.B. Wakin, IEEE Signal Process. Mag. **25**(2), 21 (2008)
24. W.L. Chan, K. Charan, D. Takhar, K.F. Kelly, R.G. Baraniuk, D.M. Mittleman, Appl. Phys. Lett. **93**(12), 121105 (2008)
25. D. Shrekenhamer, J. Montoya, S. Krishna, W.J. Padilla, Adv. Opt. Mater. **1**(12), 905 (2013)
26. H.T. Chen, W.J. Padilla, J.M.O. Zide, S.R. Bank, A.C. Gossard, A.J. Taylor, R.D. Averitt, Opt. Lett. **32**(12), 1620 (2007)
27. M.v. Exter, D. Grischkowsky, Appl. Phys. Lett. **56**(17), 1694 (1990)
28. C.C. Nadell, C.M. Watts, J.A. Montoya, S. Krishna, W.J. Padilla, Adv. Opt. Mater. **4**(1), 66 (2016)
29. C.M. Watts, C.C. Nadell, J. Montoya, S. Krishna, W.J. Padilla, Optica **3**(2), 133 (2016)
30. S. Savo, D. Shrekenhamer, W.J. Padilla, Adv. Opt. Mater. **2**(3), 275 (2014)
31. H.T. Chen, W.J. Padilla, M.J. Cich, A.K. Azad, R.D. Averitt, A.J. Taylor, Nat. Photonics **3**(3), 148 (2009)
32. S. Rout, S.R. Sonkusale, APL Photonics **1**(8), 086102 (2016). Licensed under CC BY 4.0 (http://creativecommons.org/licenses/by/4.0)
33. A. Roggenbuck, H. Schmitz, A. Deninger, I.C. Mayorga, J. Hemberger, R. Gusten, M. Gruninger, New J. Phys. **12**(4), 043017 (2010)

Chapter 6
A Terahertz Focal Plane Array Using Metamaterials in a CMOS Process

Metamaterial research has seen some unprecedented growth in the last two decades. Yet, compact and low-cost metamaterial devices are still infancy. Implementing the designs in a CMOS process will be a right step towards that goal although a huge engineering challenge. In this chapter, a method to implement a metamaterial based terahertz detector is demonstrated. A combination of electromagnetic and circuit simulations shows the viability of such a design, with the hope of serving as a reference for metamaterial designers who are interested in using the CMOS process.

6.1 Introduction

As mentioned in Chap. 5, there has been a significant interest in terahertz (THz) imaging because of its potential for wide range of applications [1–4]. Due to the difficulty in building THz focal plane array (FPA) detectors, a single-pixel imaging method was demonstrated in Chap. 5. Although compressed sensing (CS) [5, 6] techniques can be employed to speed up the frame rate, it is still a slow process due to calorimetric detection that are thermal-time-constant limited. Moreover, incoherent THz power detectors have primarily relied on specialized process technologies for the fabrication of Schottky diodes [7], or bolometers [8], making it difficult to mass produce them at lower manufacturing cost.

Complementary metal-oxide semiconductor (CMOS) process technologies are still the choice for low-cost high volume applications. They also provide the option to build highly integrable supporting circuits that allow complex electronics systems to be built on the same substrate as the sensor. Therefore it is a natural desire to implement terahertz technologies in CMOS to commercialize this emerging technology and make it available to the masses. But THz detection using CMOS has been limited to below 300 GHz due to lack of suitable low-noise amplifiers [9]. BiCMOS technologies have been developed with integration of silicon Schottky diodes with cut-off frequencies higher than 1 THz [7, 10].

S. Rout, S. Sonkusale, *Active Metamaterials*, DOI 10.1007/978-3-319-52219-7_6

Visible light digital cameras have been implemented in conventional CMOS process technologies with excellent performance [11], resulting in a massive commercialization of such cameras with 5+ Mega-pixel cameras as a standard feature even in low-cost mobile phones. However, long wavelength terahertz electromagnetic waves lack the energy for band-gap transition in CMOS technologies. Power detectors using field-effect transistors (FETs) as square law devices have been used for many years [12, 13], but have been limited to lower microwave frequencies because of low electron mobility and channel resistance, especially in CMOS technologies. In recent years, FETs have been used as terahertz detectors [14, 15], explained as non-resonant response to plasma-wave excitation of a 2D electron gas [16], but their room temperature sensitivity is still questionable for commercial application.

The first fully integrated CMOS focal plane array operating at 600 GHz was demonstrated using the principle of distributive resistive self-mixing [17, 18]. This principle allows FET power detection of frequencies well above the device cutoff frequency (f_T). The authors demonstrated the principle by implementing a 3×5 pixel CMOS FPA in a 0.25 μm NMOS technology using an on-chip patch antenna as broadband receive antenna for each pixel.

Using the same principle to detect THz waves, a focal plane array (FPA) using metamaterials as the receive antenna is demonstrated in this chapter. The use of metamaterials allows for frequency selective pixels and since the unit cells are sub-wavelength in size, multiple resonant frequency metamaterials can be packed into each pixel allowing for pseudo-color THz imaging providing spectral information of the imaged objects.

6.2 A 0.18 μm CMOS Foundry Process Technology

The decision to use a certain CMOS technology is primarily related to performance and cost. The newer and finer technologies offer significant performance advantage at higher cost. A useful frequency range for THz imaging is 230–360 GHz. A 0.18 μm CMOS foundry process is a very low-cost process due to its age and maturity. Therefore, designing a THz FPA in the frequency range of 230–360 GHz using a 0.18 μm CMOS foundry process is a very desirable device.

The conductive layer and inter-layer dielectric thickness are shown in Fig. 6.1a for a typical 0.18 μm CMOS process. It has five to six metal layers for connectivity and the top metal layer is typically a thick layer (e.g. 2 μm) which is used to design the metamaterial for smaller loss and it's the first layer after the passivation. Figure 6.1b shows the cross-section of the FET devices along with the metal layers. At the very bottom is a 500 μm thick p-type silicon substrate which is opaque to the terahertz frequencies of operation. Therefore, this technology cannot be used for any terahertz transmissive designs.

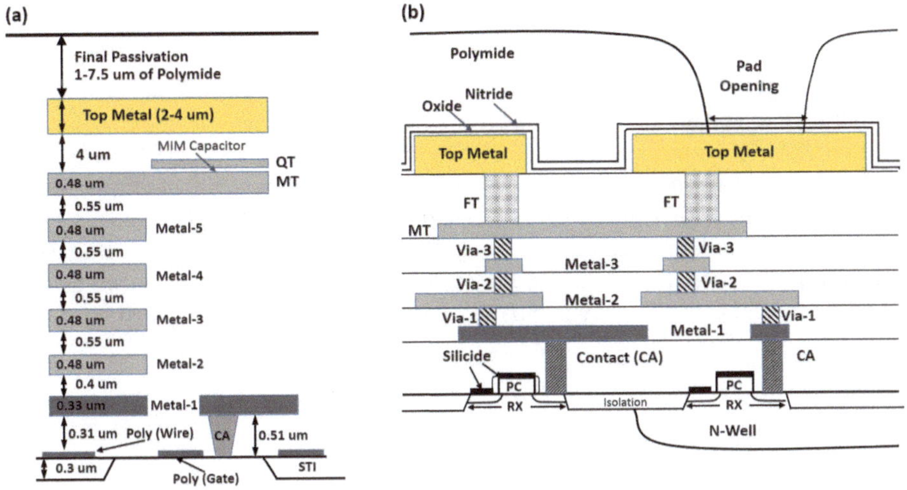

Fig. 6.1 (**a**) Conductive film and inter-layer dielectric thickness for a 0.18 μm CMOS foundry process. (**b**) *Cross-section view* of the layers with FET. (Not drawn to scale)

Table 6.1 Properties of the CMOS materials used for electromagnetic simulations

Material	Unit	Value
p-Si substrate relative permitivity		11.9
p-Si substrate conductivity	S/m	7.41
$Si0_2$ Conductivity	S/m	0.00
$Si0_2$ relative permitivity		4.10
$Si0_2$ Dielectric loss tangent		0.00–0.001
Polymide relative permitivity		3.4
Polymide loss tangent		0.00–0.001
All dielectrics relative permeability		1.00

NFETs are formed in the p-type substrate within a p-well. For isolating active devices, shallow trench isolation (STI) is used everywhere except the active areas (RX in Fig. 6.1b). From various flavors of FETs available, a minimum length and width standard NFET device should be used for the terahertz detection. The standard NFET device operates at 1.8 V, has an effective gate thickness of 3.5 nm ($C_{\mathrm{ox}} = 7.9\,\mathrm{fF}/\mu\mathrm{m}^2$), minimum drawn length of 0.18 μm, threshold voltage $V_{\mathrm{th}} \approx 355\,\mathrm{mV}$ and mobility $\mu_n \approx 400\,\mathrm{cm}^2/\mathrm{V}\cdot\mathrm{s}$.

The electromagnetic properties of the CMOS materials for simulating the metamaterials are tabulated in Table 6.1. In a foundry process, all the vertical dimensions are fixed in a given process and design rules are set for minimum width and spacing for metal and active layers, among numerous other design rules. Most foundry processes use chemo-mechanical polishing (CMP) to flatten uneven oxide layer after each metal and oxide deposition. This imposes design rules on having minimum density for each metal layers which is typically about 30% for all the metal layers. This introduces challenges in metamaterial designs since we cannot have empty areas underneath the metamaterials.

6.3 Principle of Resistive Self-Mixing Detection

In order to get an intuitive feel of the resistive self-mixing principle, the reader is encouraged to see the quasi-static analysis of FET square law detector by Ojefors et al. in [18].

For input frequencies well above the transit frequency, the non-quasi-static (NQS) analysis with a distributed RC model of the FET channel needs to be considered as shown in Fig. 6.2. Each segment resembles a quasi-static self-mixing detector [18], where the transistor is replaced by the segment conductance $g_{n-1}(v)$ and the external $C_{gd,ext}$ is provided by C_n. Kirchoff's junction rule at node n yields

$$i_{g,n-1} - i_{g,n} = i_{C,n} \tag{6.1}$$

This equation is equivalent to

$$g_{n-1}(v_{n-1} - v_n) - g_n(v_n - v_{n+1}) = C_n \frac{d}{dt}(v_n) \tag{6.2}$$

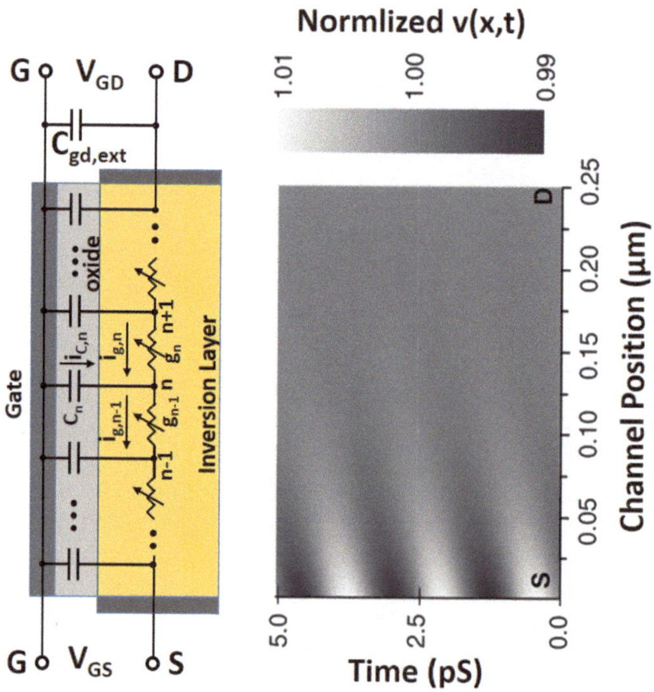

Fig. 6.2 Non Quasi-static (NQS) analysis of self-mixing. The NQS model with distributed gate-channel capacitance C_n is shown on the *left*. On the *right*, numerical solution of the time and position dependent gate-to-channel voltage $v(x, t)$ described by Eq. (6.5) for a 0.25 µm device excited at the source-gate junction with a small 600 GHz signal is shown [18]

where each conductance g_n can be further expressed by a conductivity per unit length $G(v(x,t))$, which depends on the local gate-to-channel voltage $v(x,t)$ and the segment length Δx as follows:

$$g_n = \frac{G(v(n\Delta x, t))}{\Delta x} \tag{6.3}$$

Similarly, the individual segment capacitors can be expressed as

$$C_n = C_{\text{ox}} W \Delta x \tag{6.4}$$

where C_{ox} is the gate capacitance per unit area and W is the width of the device. After combining Eqs. (6.2)–(6.4), and reducing the difference equation to a partial differential equation, for strong inversion it can be shown as [18]

$$\frac{\partial}{\partial x}\left[\mu\left(v(x,t) - V_{\text{th}}\right)\frac{\partial v(x,t)}{\partial x}\right] = \frac{\partial}{\partial t} v(x,t) \tag{6.5}$$

It can be seen from Eq. (6.5), the NQS response is only dependent on the carrier mobility μ and the bias voltage $V_g - V_{\text{th}}$. The NQS self-mixing response can be obtained by solving Eq. (6.5) with the following boundary conditions:

$$v(0,t) = V_{\text{RF}} \sin \omega t + V_g \tag{6.6}$$

$$v(L,t) = V_g \tag{6.7}$$

Figure 6.2 shows a 600 GHz numerical solution of Eq. (6.5) for the normalized gate-to-channel voltage $v(x,t)$ in a 0.25 μm NMOS device as reported in [18]. The initially applied voltage propagates through the channel from the source (left) towards the drain (right) while it exhibits an exponential damping. After 0.1 μm, the signal is heavily attenuated, and the channel has essentially reached the drain potential. Hence, efficient resistive mixing takes place close to the source while the rest of the device acts as distributed capacitance C_{gd} and parasitic series resistance. This allows a long-channel device to be used for direct power detection even at terahertz frequencies. Distributed resistive self-mixing confines the mixing action to a much smaller section of the transistor and charges do not need to propagate through the entire length of the device. However, it is expected that the use of a shorter device will improve the detection performance as the thermal noise contribution from the non-modulated part of the channel is reduced. A shorter device implemented in a more advanced process technology is also likely to have smaller parasitic substrate capacitances, thus increasing the responsivity.

6.4 Metamaterial Based Terahertz CMOS Detector Design

6.4.1 Terahertz Detection Using Source-Driven Self-Mixing Architecture

In the previous section we described the terahertz detection principle using a gate and drain driven circuit topology. The same result can be achieved by driving the source end, with the gate and drain as virtual ground. The advantage of this method is its broadband nature and architecturally suits well for terahertz detection with metamaterials.

Figure 6.3 shows a simplified circuit schematic of a terahertz detector using source-driven distributive self-mixing. In contrast to the gate-drain coupling approach used in the previous section, the RF power from the metamaterial is provided to the source terminals of the two NMOS transistors NM1 and NM2 in the present design. The gate and drain terminals of the transistors are connected to DC bias voltages, thus creating virtual grounds for the RF voltage. In a balanced configuration, half of the terahertz signal generated by the metamaterial split gap appears as a voltage across the gate-source junctions (V_{GS}) of each transistor and contributes to the distributed self-mixing process in the channel. The rectified output current is extracted from the shared drain node and integrated on an on-chip capacitor which is compared with a reference using a differential amplifier, serving as a comparator, to indicate terahertz detection. This configuration eliminates the need for quarter-wave stubs and coupling capacitors, which are necessary in the gate-driven detector design [18] in order to tie the RF potential of the gate and drain together and to provide isolation of the output port from the antenna. Hence, a wider operating bandwidth can be obtained with the source-driven detector than with the gate-driven one.

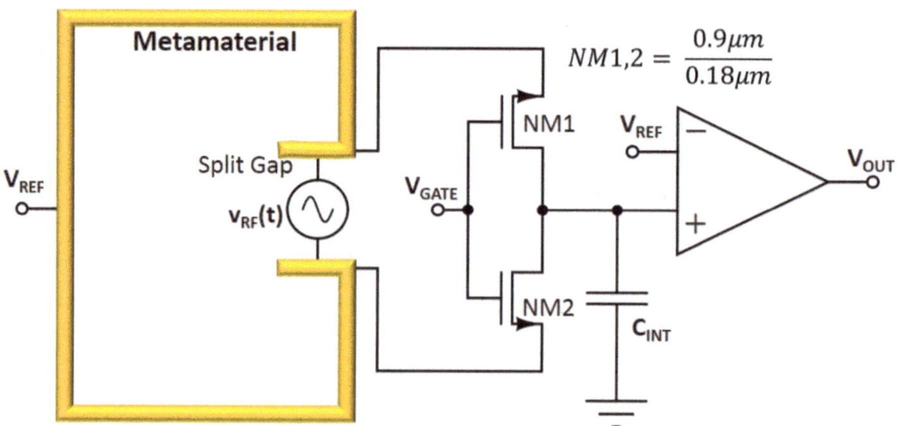

Fig. 6.3 Schematic of the differential source-driven distributed-mixing terahertz square-law power detector using n-channel MOS transistors in a 0.18 μm CMOS process

6.4.2 Circuit Architecture for Terahertz Detection

The architecture for terahertz detection of each pixel is based on the source-driven self-mixing principle as described in the previous Sect. 6.4.1. Figure 6.4 shows a simplified schematic of the implemented architecture in 0.18 μm CMOS technology. The core of the THz detector circuit is a series connected NMOS pair, NM1 and NM2, with their sources connected to the split gap of a metamaterial unit cell in each pixel.

During the *reset phase* (ϕ is high), the integration cap C_{int} and the gate of PM1 and PM2 are connected to the bias voltage V_{ref}. Similarly, the gate of PM2 is connected to V_{ref} through the dummy transistors NM3 and NM4. This sets an equal currents in the differential stage branches PM1 and PM2. The dummy replica stage assures that the diff-pair reference mimics any switching effects associated with the detector NMOS pair NM1 and NM2. Please note, any random or systematic mismatch in PM1 and PM2 can be corrected by adjusting one of the V_{ref} voltages.

During the *detection phase* (ϕ is low), the metamaterial is disconnected from the bias voltage and the terahertz electric field generated in the split gap will generate a DC current in NM1 and NM2 that is integrated on C_{int}. This integrated voltage creates a differential current in the differential pair (diff-pair), PM1 and PM2, since the reference side of the diff-pair (gate of PM2) should hold constant. The differential current generated from the diff-pair is fed into a folded cascode stage (PM3-PM6, NM6-NM9) [19] with a combined class-AB output stage (PM7, PM8, NM5, NM10) that converts the differential current to voltage output. The purpose of the class-AM output stage is to be able to drive small resistive load.

Finally, the output of the folded cascode stage will be a periodic signal with the same periodicity as ϕ, with the pulse width proportional to the incident THz power which is shown in a simplified timing diagram in the lower-left inset of Fig. 6.4.

6.5 Metamaterial Design for Terahertz Detection

As detailed in Sect. 6.4.1, the method for THz detection is to use the concentrated electric field in a metamaterial split gap to create a DC current using a pair of source-driven NFETs. So the metamaterials design will primarily focus on maximizing the electric field in the split gap. The first experimental demonstration of metamaterial perfect absorber (MPA) in 2008 [20] used split ring and a cut wire underneath to absorb EM radiation at a desired frequency. It was shown in that work that most of the absorbed energy was in dielectric losses occurring in between the two metamaterial elements where the electric field is large. Unfortunately, that is not an ideal location for THz detection in planar semiconductor technology.

Since the goal is to maximize the electric field at the split gap, electrical-LC (ELC) resonator, shown in Fig. 6.5a, is a natural choice. The unit size is 70 μm and repeated every 120 μm. The pair of NFET detector is placed underneath one of the

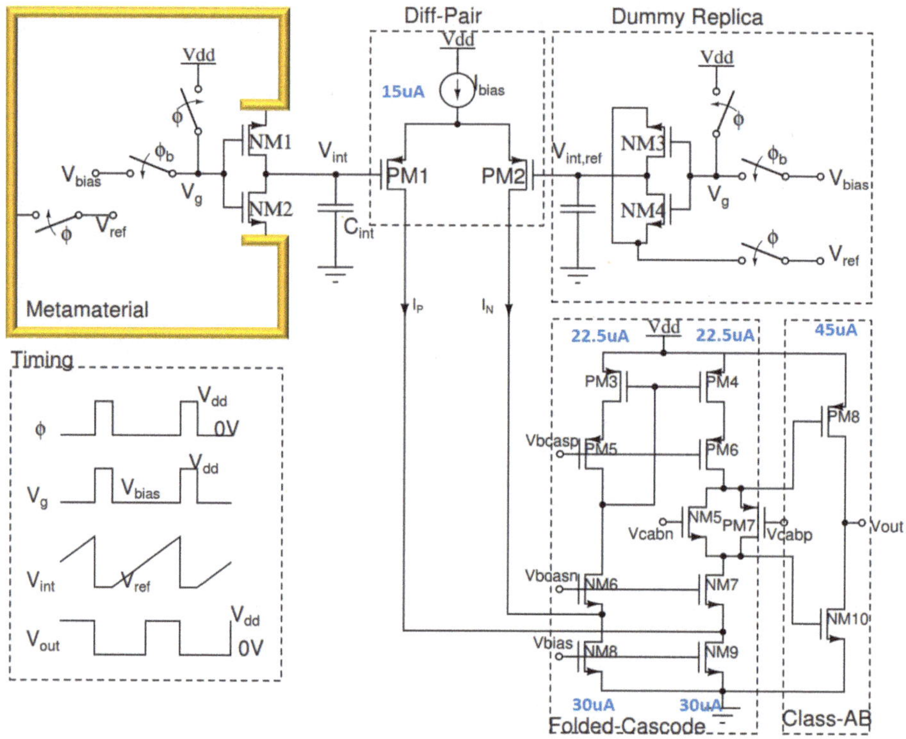

Fig. 6.4 Circuit architecture for terahertz detection. A pair of n-channel MOS transistors (NM1 and NM2) are connected between the metamaterial split gap in the source-driven configuration. During the reset phase (ϕ is high), the integration cap C_{int} and the gate of PM1 and PM2 are connected to the bias voltage V_{ref}. During the detection phase (ϕ is low), the metamaterial is disconnected from the bias voltage and the electric field generated in the split gap will generate a DC current in NM1 and NM2 that is integrated on C_{int}. A dummy replica is used to generate the reference for the diff-pair and mimic all the switching effects of NM1, NM2. The differential current generated from the differential pair (diff-pair) PM1 and PM2 is fed into a folded cascode stage that converts the differential current to voltage output. And the folded cascode is followed by a class-AB output stage to drive small resistive load. A simplified timing signal with associated voltage is shown in lower left inset. DC bias currents for the main branches are shown in the figure and all transistor sizes are tabulated in Table 6.2

split gaps and the drains of the NFETs are connected to the split gap with a set of vertical vias as shown in the inset. For electromagnetic simulations, a discrete port with an impedance of 300 Ω at the end of the vias is used to represent the lumped circuit equivalent of the NFET detector. Figure 6.5b–d shows the resonant field and current distributions at the metamaterial plane at 325 GHz. The electric field plot in Fig. 6.5c shows the enhanced electric field in the split gap and the asymmetry is due to the load on only one of the split gaps. The magnetic field plot shown in Fig. 6.5d

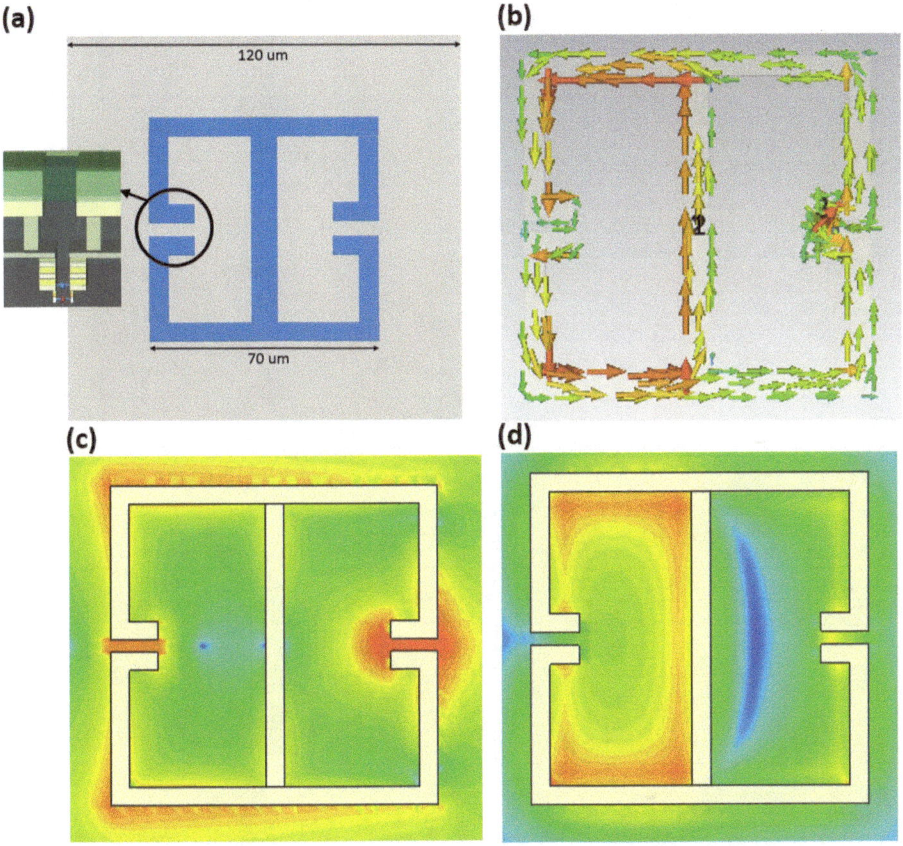

Fig. 6.5 Metamaterial design and EM simulation results. (**a**) ELC with unit size of 70 μm designed to resonate at 325 GHz. The NFET detector is placed underneath one of the split gaps with the drains of the NFETs connected to the split gap through a set of vias as shown in the inset. (**b**)–(**d**) Resonant field and current distribution at the metamaterial plane: (**b**) surface current distribution, (**c**) electric field distribution, and (**d**) magnetic field distribution

also shows that the field in both the halfs do not cancel each other because of the asymmetric load. That is not an issue with this design since we are not trying to create an effective material.

The key transfer function for this detector is the voltage output of the discrete port for a given incident THz power in a unit cell of $120 \times 120\ \mu m^2$. Figure 6.6b shows a peak of 9.27 V/W at 325 GHz. The S11 reflection plot in Fig. 6.6a shows that the peaks for the effective medium is not necessarily same as that of peak THz detection.

Table 6.2 Transistor sizes for the terahertz detector circuit in Fig. 6.4

Transistor	W(μm)/L(μm)
NM1-4	0.9/0.18
NM5	10/1.2
NM6,7	10/2
NM8,9	10/4
NM10	30/0.18
PM1,2	10/0.18
PM3,4	5/2
PM5,6	20/0.25
PM7	10/0.25
PM8	60/0.18

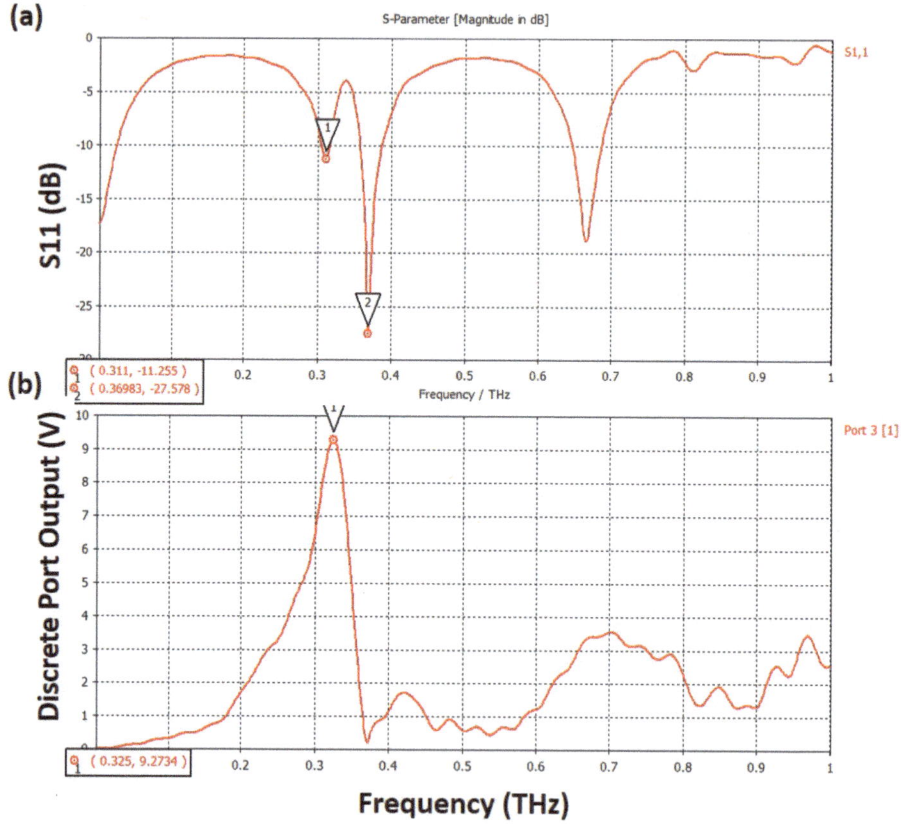

Fig. 6.6 (**a**) The reflection co-efficient S11 plot. (**b**) Voltage at the discrete port for 1 W of incident power on each unit cell of 120 × 120 μm²

6.6 Design of the Test Chip in 0.18 μm CMOS Process

The layout of the test chip is shown in Fig. 6.7. The test chip contains four different configuration of metamaterials, two different types of resonators (ELC and SRRs) and for each resonator, two different frequencies as tabulated in Table 6.3.

Each configuration consists of a 3 × 2 array of THz metamaterial detectors surrounded by passive metamaterials to minimize scattering effect. The diff-pair output of each element of the 3 × 2 array is connected together and fed to a folded cascode stage, one for each configuration. The digital circuitry controls which element is chosen and the diff-pair output current of that selected element is connected to the folded cascode output stage to generate output. The digital block contains a set 16 registers that are used to select different metamaterial elements and some modes for troubleshooting. The registers are written using a shift register with a standard SPI protocol. A common bias block provides bias currents to all the blocks.

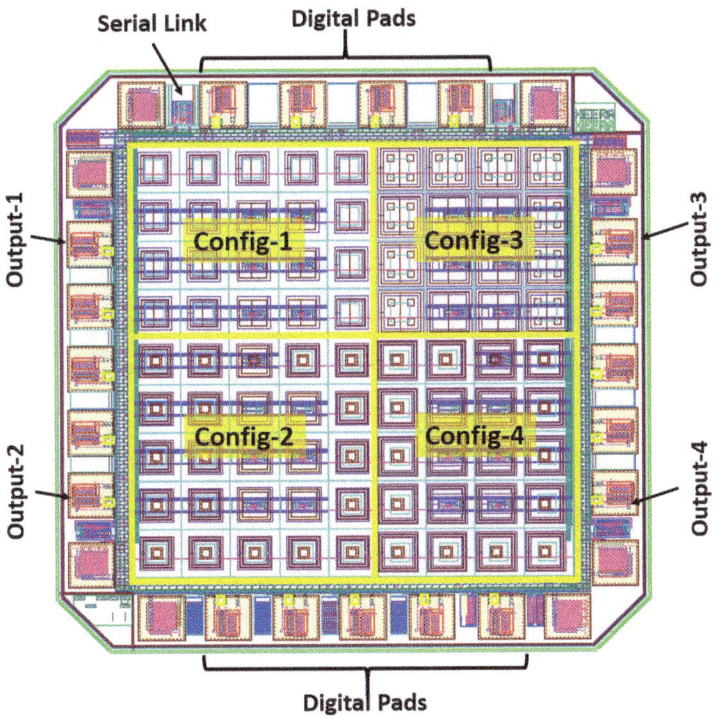

Fig. 6.7 Annotated image of the layout of the complete test chip in the 0.18 μm CMOS process

Table 6.3 Design parameters of the four different configurations in the test chip

Configuration	Resonator	Unit size (μm)	Frequency (GHz)
Config-1	SRR	80 × 80	320
Config-2	SRR	96 × 96	270
Config-3	ELC	70 × 70	325
Config-4	ELC	100 × 100	230

6.7 Circuit Simulation Results

In order to do a complete simulation, the THz voltage at the NFET detectors can be estimated at the metamaterials split gap from the simulation results shown in Fig. 6.6a for a known incident power. And that THz voltage can be applied in the circuit simulations provided non-quasi-static (NQS) model is available for the FETs. When NQS models are not available, the DC current can be predicted from analysis or numerical simulation and that current is applied in the circuit simulations. For a 1 mW THz source illuminating about 2 mm^2 of the detector area, about 1–5 pA of detector current is estimated for this design.

All the circuits were simulated using the industry standard BSIM3 models provided by the foundry. Figure 6.8 shows the simulation results for a single detector circuit. Figure 6.8a shows the reset and detect pulse with a pulse width of 2 μs and a period of 200 μs. Figure 6.8b shows the voltages on the two inputs of the differential pair, one is the detector integrating voltage and the other is a dummy replica. It can seen from the plot that both voltages are reset to 500 mV and when the switches are turned off/on for the detection phase, there is about 100 mV switching feedthrough and it can also be seen that both the integrating capacitors discharging strongly due to leakage currents of the parasitic diodes of the NFETs. That is the primary reason for creating the dummy replica stage to cancel the common non-linear effects. The difference of the voltages are plotted in Fig. 6.8c for a 1 pA detection current. Figure 6.8d shows the final output of the folded cascode output stage with a load of 100 kΩ resistor in parallel with 10 pF capacitor. The plot shows result of two detector currents, 1 pA (blue) and 2 pA (red). The method of measuring the difference in the incident THz power is to measure the difference between the pulse widths and back calculate the incident THz power.

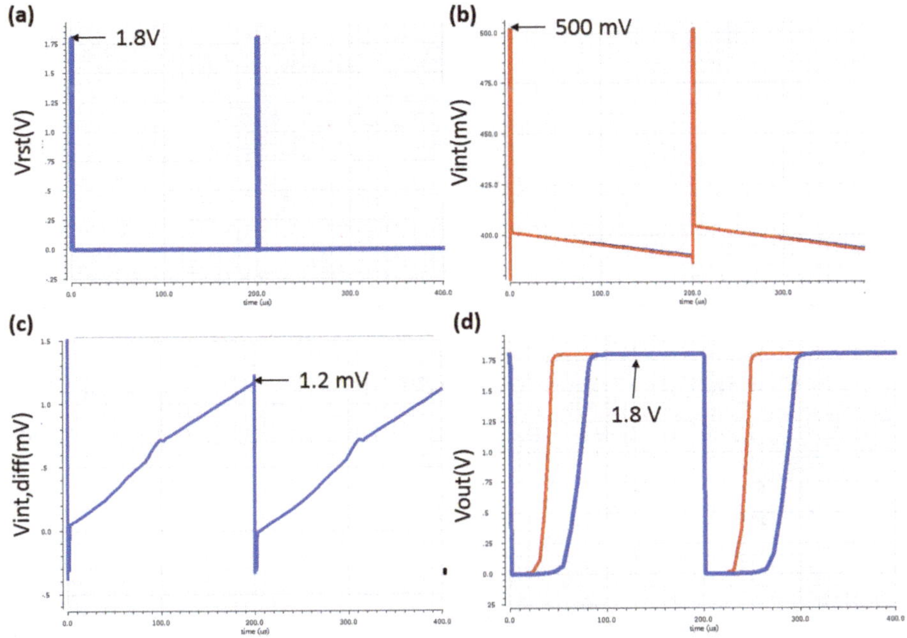

Fig. 6.8 Circuit simulation results. (**a**) The reset and detect signal with reset width of 2 μs and period of 200 μs. (**b**) Integrating voltages on both sides of the differential pair with 1 pA current injected on the metamaterial side. (**c**) Differential voltage of the two integrated voltages. (**d**) Output voltage of the folded cascode for two different detection current 1 pA and 2 pA [21]

References

1. T.M. Korter, D.F. Plusquellic, Chem. Phys. Lett. **385**(1–2), 45 (2004).
2. N. Karpowicz, H. Zhong, C. Zhang, K.I. Lin, J.S. Hwang, J. Xu, X.C. Zhang, Appl. Phys. Lett. **86**(5), 054105 (2005)
3. K. Kawase, Y. Ogawa, Y. Watanabe, H. Inoue, Opt. Express **11**(20), 2549 (2003)
4. R.M. Woodward, V.P. Wallace, R.J. Pye, B.E. Cole, D.D. Arnone, E.H. Linfield, M. Pepper, J. Invest. Dermatol. **120**(1), 72 (2003)
5. D. Donoho, IEEE Trans. Inf. Theory **52**(4), 1289 (2006)
6. E. Candes, J. Romberg, T. Tao, IEEE Trans. Inf. Theory **52**(2), 489 (2006)
7. J.L. Hesler, T. Crowe, Responsivity and noise measurements of zero-bias Schottky diode detectors, in *Proceedings of the 18th International Symposium on Space Terahertz Technology* (Pasadena, 2007), pp. 89–92
8. P. Helisto et al., Antenna-coupled microbolometers for passive THz direct detection imaging arrays, in *Proceedings of the 1st European Microwave Integrated Circuits Conference*, (IEEE, 2006), pp. 35–38
9. A. Tessmann, I. Kallfass, A. Leuther, H. Massler, M. Schlechtweg, O. Ambacher, in *Microwave Integrated Circuit Conference, 2008. EuMIC 2008. European* (2008), pp. 210–213
10. S. Sankaran, K.K. O, IEEE Electron Device Lett. **26**(7), 492 (2005)
11. A. Theuwissen, in *Solid State Circuits Conference, 2007. ESSCIRC 2007. 33rd European* (2007), pp. 21–27
12. R.A. Barret, Broadband RF detector using FET. Patent no 4647848 (1987)

13. H-G. Krekels, B. Schiek, E. Menzel, Power detector with GaAs field effect transistors, in *22nd European Microwave Conference*, vol. 1 (IEEE, 1992), pp. 174–179
14. W. Knap, F. Teppe, Y. Meziani, N. Dyakonova, J. Lusakowski, F. Boeuf, T. Skotnicki, D. Maude, S. Rumyantsev, M.S. Shur, Appl. Phys. Lett. **85**(4), 675 (2004)
15. R. Tauk, F. Teppe, S. Boubanga, D. Coquillat, W. Knap, Y.M. Meziani, C. Gallon, F. Boeuf, T. Skotnicki, C. Fenouillet-Beranger, D.K. Maude, S. Rumyantsev, M.S. Shur, Appl. Phys. Lett. **89**(25), 253511 (2006)
16. M. Dyakonov, M. Shur, IEEE Trans. Electron Devices **43**(3), 380 (1996)
17. U. Pfeiffer, E. Ojefors, in *Solid-State Circuits Conference, 2008. ESSCIRC 2008. 34th European* (2008), pp. 110–113
18. E. Ojefors, U. Pfeiffer, A. Lisauskas, H. Roskos, IEEE J. Solid-State Circuits **44**(7), 1968 (2009)
19. P.R. Gray, P.J. Hurst, S.H. Lewis, R.G. Meyer, *Analysis and Design of Analog Integrated Circuits*, 5th edn. (Wiley, New York, 2009)
20. N.I. Landy, S. Sajuyigbe, J.J. Mock, D.R. Smith, W.J. Padilla, Phys. Rev. Lett. **100**(20), 207402 (2008)
21. S. Rout, Active metamaterials for terahertz communication and imaging. Ph.D., Tufts University (2016)

Appendix A
Electromagnetic Waves

A.1 Helmholtz's Equation

Taking the curl of $\nabla \times \mathbf{E} + \frac{\partial \mathbf{B}}{\partial t} = 0$ in Eq. (2.4):

$$\begin{aligned}\nabla \times (\nabla \times \mathbf{E}) &= -\nabla \times \left(\frac{\partial \mathbf{B}}{\partial t}\right) \\ \nabla(\nabla \cdot \mathbf{E}) - \nabla^2 \mathbf{E} &= -\mu\epsilon \frac{\partial^2 \mathbf{E}}{\partial t} \\ \nabla^2 \mathbf{E} + \mu\epsilon \frac{\partial^2 \mathbf{E}}{\partial t} &= 0\end{aligned} \tag{A.1}$$

Similarly, the other part of the Helmholtz's equation can be solved by taking curl of $\nabla \times \mathbf{H} - \frac{\partial \mathbf{D}}{\partial t} = 0$ in Eq. (2.4).

A.2 Electromagnetic Waves *Are* Transverse

Let us consider a plane wave (in x-direction) that satisfies both Maxwell's and Helmholtz's equation

$$\begin{aligned}\mathbf{E}(\mathbf{x}, t) &= \mathbf{E}_0 e^{i(kx - \omega t)} \\ \mathbf{B}(\mathbf{x}, t) &= \mathbf{B}_0 e^{i(kx - \omega t)}\end{aligned} \tag{A.2}$$

S. Rout, S. Sonkusale, *Active Metamaterials*, DOI 10.1007/978-3-319-52219-7

Whereas every solution to Maxwell's equations (in empty space or nonconductors) must obey the wave equation, the converse is *not* true; it imposes special constraints on $\mathbf{E}_0$ and $\mathbf{B}_0$. Let us consider $\mathbf{E}(\mathbf{x}, t)$ with three orthogonal components as

$$\mathbf{E}(\mathbf{x}, t) = \left(\mathbf{E}_{0x} + \mathbf{E}_{0y} + \mathbf{E}_{0z}\right) e^{i(kx-\omega t)} \tag{A.3}$$

Since $\nabla \cdot \mathbf{E} = 0$,

$$\frac{\partial \mathbf{E}_{0x} e^{ikx}}{\partial x} + \frac{\partial \mathbf{E}_{0y} e^{ikx}}{\partial y} + \frac{\partial \mathbf{E}_{0z} e^{ikx}}{\partial z} = 0 \tag{A.4}$$

Equation (A.4) is satisfied only if

$$\mathbf{E}_{0x} = 0 \tag{A.5}$$

Similarly, it can be shown that

$$\mathbf{B}_{0x} = 0 \tag{A.6}$$

Moreover, it can be shown from Faraday's law, $\nabla \times \mathbf{E} = -\partial \mathbf{B}/\partial t$ that

$$\mathbf{B}_0 = \frac{k}{\omega}\left(\hat{i} \times \mathbf{E}_0\right) \tag{A.7}$$

Thus, Eqs. (A.5)–(A.7) show that the EM plane wave needs to be *transverse* with the vectors $\mathbf{B}_0, \mathbf{E}_0, k$ forming a right-handed triplet.

FSC
www.fsc.org
MIX
Papier | Fördert gute Waldnutzung
FSC® C083411

Kolibri 360 GmbH
Ferdinand-Jühlke-Straße 7
99095 Erfurt, Deutschland
produktsicherheit@kolibri360.de